NARROW PODCASTING

Make profitable connections
and grow your business, without paid
ads, sponsors, or thousands of listeners

NARROW PODCASTING

Make profitable connections
and grow your business, without paid
ads, sponsors, or thousands of listeners

TOBY GOODMAN

Copyright © 2022 by Toby Goodman

All rights reserved. No part of this book may be used or reproduced in any manner whatsoever without prior written consent of the authors, except as provided by the United States of America copyright law.

Printed in the United States of America.
ISBN:978-1-956649-38-3

This publication is designed to provide accurate and authoritative information with regard to the subject matter covered. It is sold with the understanding that the publisher is not engaged in rendering legal, accounting, or other professional advice. If legal advice or other expert assistance is required, the services of a competent professional should be sought.

For more information, please email:
Toby Goodman at hello@narrowpodcasting.com

TABLE OF CONTENTS

WELCOME TO THE WORLD OF NARROWCASTING 1

CHAPTER 1: POD POWER. 7
    WHY I LOVE A POD, AND YOU SHOULD TOO 8
    WHAT'S HOLDING YOU BACK? 10

PREPRODUCTION: TAKE AIM 15

CHAPTER 2: NEVER START A PODCAST WITHOUT THESE. 17
    THREE PRE-PODCAST MUST-HAVES 18
    PODCAST POSITIONING. 22

CHAPTER 3: MIND YOUR BUSINESS (MAKING PODCASTING WORK) 29
    GOOD, BETTER, BEST 31
    ARE YOU A ONE-MAN BAND? 32

CHAPTER 4: CUTTING THROUGH THE NOISE 35
    1. KEEP IT SHORT 'N' SWEET 36
    2. EYE-CATCHING TITLES AND GUESTS. 37
    3. LAUNCH STRATEGY: THE '10' GAMBIT. 38
    4. CREATE A TRAILER 39
    5. GET PERSONAL 39

CHAPTER 5: FIVE LITTLE POD DUCKS 41
- DUCK ONE: BUILD A HOME FOR YOUR PODCAST ONLINE 41
- DUCK TWO: CREATE YOUR PIPELINE 43
- DUCK THREE: TECH BASICS 44
- DUCK FOUR: LEGAL BASICS 46
- DUCK FIVE: INTRO/OUTRO 47

PRODUCTION: HITTING THE TARGET 51

CHAPTER 6: MIC DROP: PRESENT LIKE A PRO 53
- EPISODE SCRIPT 54
- TONE AND EMOTION SCRIPT 58
- SOUND QUALITY: THE TECHIE BIT 59

CHAPTER 7: BE MY GUEST (GREAT INTERVIEW SHOWS) 63
- MEET THE POD-LATIONSHIP CYCLE 64
- THE RIGHT INVITE 67
- MENTIONITIS 69
- IT'S A YES! SO, WHAT'S NEXT? 71
- PODCAST PREP FOR GUESTS 72
- TECH PREP WITH GUESTS 76
- ZOOM RECORDING: MEET THE TWO UNRELATED ZOOMS! 77
- A FINAL RECORDING NOTE 78
- RECORDING SUPER-HIGH VALUE GUESTS 79
- WHEN THE TABLES ARE TURNED 81

CHAPTER 8: EDITING FOR SUCCESS 83
- INCLUSIVE INTERVIEWING 84
- BE PREPARED 85
- BUILDING YOUR PODCAST 86
- 20-SECOND TEASER 87
- WHAT DO YOU DO WHEN YOU'VE TOO MUCH GOOD CONTENT FOR ONE SHOW? 88
- WHAT IF YOU MISS A RECORDING/PUBLISHING DEADLINE? 88

FREQUENCY AND CADENCE 89

MORE WAYS TO GET VALUE FROM YOUR EPISODES 90

POSTPRODUCTION: PROMO 93

CHAPTER 9: GETTING SHOWS OUT THERE 97

PERFECT PODCAST PROMO 98

SEARCH ENGINE OPTIMIZATION: DON'T FORGET YOUTUBE! 101

ACCESSIBILITY: MORE ON INCLUSIVE PODCASTING 101

OWNERSHIP: AN IMPORTANT REMINDER 102

CREATE SHOW NOTES 103

PODCAST NETWORKS 106

CHAPTER 10: AMP UP YOUR NETWORK (GROWING YOUR AUTHORITY) 109

STAY TOP OF MIND: THE FOLLOW-UP 112

ALUMNI CONNECTION 117

MATCHMAKING 120

USE YOUR LEVERAGE 121

POD-LATIONSHIP CYCLE IN ACTION 122

TRACKING SOLO SHOWS 125

CHAPTER 11: SHOW ME THE MONEY (MORE WAYS OF GENERATING REVENUE) 129

PERMISSION-BASED MARKETING 130

WHAT TO CHARGE 132

CHAPTER 12: DON'T BE LIKE DAVE 135

ULTIMATE PROFITABLE POD CHECKLIST 137

FROM POD-CURIOUS TO PROFITABLE POD-LATIONSHIPS 141

THANKS AND ACKNOWLEDGMENTS 143

ABOUT THE AUTHOR 145

WELCOME TO THE WORLD OF NARROWCASTING

Narrowcast (ˈnærəʊˌkɑːst)[1]
vb, **-casts**, **-casting**, **-cast** *or* **-casted**
1. (Broadcasting) (*tr*) to supply (television programmes) to a small area by cable television
2. (Broadcasting) (*intr*) (of programmers or advertisers) to target a specialized audience on radio or television

I don't imagine you think much about airport flight information screens until you're about to fly. Then, you check them every three minutes until they tell you your gate number. These screens are a perfect example of narrowcasting: they display flight information only applicable to passengers leaving from that airport on that specific day—a very narrow set of ticket holders.

In this book, I will show you how to create a narrow podcast, because unlike one of those screens, your podcast can be even more personal—as if the screen mentioned each passenger by name. This method is so laser targeted, it'll move you away from concerning yourself with the vanity metrics of unidentified listeners and toward real connections with specific people that lead to sustained and profitable business growth.

---

1 www.thefreedictionary.com/narrowcast

Once an incredibly narrow medium, podcasting is now a widely enjoyed part of the traditional broadcast media landscape. Podcasts are so ubiquitous, it's understandable that business owners ask if podcasting is worth the hassle. After all, as the broad appeal widens, there's less chance than ever you'll grow a podcast through listenership and sponsorship. And when it comes to paid advertising online, the whole exercise feels more like a risk than a savvy investment.

Maybe you're a bit like me. You view social media as a necessary evil and would much rather connect with someone through an actual conversation. Also, perhaps you prefer helping others before yourself, even though, as a business owner, you've been told to fit the oxygen mask to your own business before it suffocates. I understand that podcasting might feel like the wrong medium for you. Indeed, most business owners get podcasting badly wrong. They lose time and money on an ill-defined shiny pod, and the distraction costs them big time. But not you—not now you have this book.

Be prepared to suspend your preconceived ideas about what podcasting is, and get ready to discover exactly how, by going narrow, it can serve you. I'm NOT going to teach you how to grow your podcast to get more listeners. Instead, I'm going to show you exactly how I've helped my clients grow their businesses using what I call the **Profitable-Pod Method**. The method teaches you that if you are focused enough, you will see a new path, one so narrow, it only has room for you and the people that (when it comes to growing your business) truly matter.

For the last five years, I've helped independent business owners all over the world implement this method to create podcasts that help them on their mission to create meaning, authority, and profit. When you understand and implement the system properly, I guarantee you'll never need to look at a podcast stat again, worry about how to measure the success of your show, or get wrapped up in the increasingly broad podcasting echo chamber of narcissists, attention seekers, and noise makers. And, if you're running a business but don't consider yourself an extrovert, then narrow podcasting is going to be perfect for you.

The beginnings of the Profitable-Pod Method started in 1999 (though I didn't know that at the time) because that's when I got my first well-paid, long-term gig in an unexpected way. Despite the fact podcasts didn't exist then, the skills I came to acquire back then were the foundations of the method.

As a career drummer, my dream was never to be famous; it was simply to be able to play at the highest level with the best artists and get paid for doing it. To make this happen, I needed to be consistently good, easy to work with, as well as known, liked, and trusted by the right people.

I was 18, studying music at college in London, and keen to be a professional musician. Music was all I'd ever wanted to do, but my degree course was technical and didn't cover how to actually get a job. Furthermore, to be seen looking for work was (and still is) deemed uncool. If you're proactively looking, you're either thought of as a hustler (a negative word in the music business) or, at worst, thought of as desperate (in the same way that if you have to ask how much a suit is on Savile Row, you probably can't afford it). But then, in the final weeks of my first year, I picked up a wedding gig from some more experienced musicians I'd got in with at a jam session.

This was a fairly well-to-do crowd, and the wealthy people of north London ate their food while my band mates and I provided soft ballads and swinging classics. In the break between sets, I picked up my Nokia phone. Just holding it made me feel like a professional. Music producer Quincy Jones still hadn't called, but I did have an unread text message. It was just my old friend Shelly, who I knew from the school bus. The text read: Free house. Got some weed. Tom, Andy and Hannah here. Come over if you've got enough petrol. At around 11:30 p.m., I walked into the cloud of smoke that filled Shelly's living room.

"This is my Auntie Sue," Shelly said. Sue looked up from her wine glass and said, "Alright, Toby! What are you wearing that for?" In my haste to get away from the posh drunk people, I'd not removed my dickie bow. I proudly announced I was a freelance drummer and had just come from a gig.

"Drummer?" she said.

I nodded.

"I'm an agent," said Sue.

Bullshit, I thought. *You're Shelly's drunk aunt.* Then she handed me a business card. It impressed me. No one had handed me a business card before. It read:

STEVEN MICHAELS MUSIC
SUE FOSTER - AGENT.

It had her phone number and address. *That looks proper and professional*, I thought, the smoke helpfully disguising my excitement.

"Send me your stuff next week, yeah?" said Sue, as she went back to nursing her glass of red.

On Saturday, I woke up in my student halls and wondered what I should do. I put a resume together detailing the youth bands, original (unheard of) bands, where I was at college, and a list of venues I had played in. Then I drove across London to her house to deliver it in person.

Nervously, I knocked, and a sober and smartly dressed Sue answered. Presenting the envelope as coolly as possible, I asked if there was anything she could use me for; I'd love to be considered. Sue's office was covered in posters of pop stars of the day, from Take That to Billie Piper, with her desk littered with headed notepaper from the major record labels and TV networks.

But despite feeling like a tiny fish in her big showbiz pond, Sue seemed genuinely impressed that I got my flimsy resume to her so quickly. A couple of days later, the phone rang. I'd been put up for a pop video audition, which turned out to be for a reggae band. Turns out, I wasn't a great physical fit, but I kept in touch with Sue, and as she requested, I sent her a few other musicians who were on my course.

Finally, the holidays came, and I was back in my room at my dad's house. One morning in late August, my Nokia lit up with a voice message. Sue Foster had "highly recommended" me for nine weeks' work at the 3000-seat Blackpool Opera House. The fee, £600 a week. At that time, I was probably earning £35 a week, so to be offered so much was insane to me. Realizing I needed to defer my second year of university, I convinced my dad to let me take a year off so I could see what it was like to be a working musician. I drove to Blackpool … and the rest is history.

That meeting with Sue, and my ability to capitalize on it, was to become the essence of Profitable-Pod because it's all about making a personal connection. Had I "broadcast" 1000 CVs to kick-start my career, I would not have had the success I attained with my "narrowcast" method of getting in there and following through with Sue.

In the 20 or so years that followed, I traveled to 25 countries and played thousands of West End and theater shows. I've been hired to play for a Rolling Stone, a Bee Gee, a Spice Girl, and more TV talent show competition winners (and losers) than the public can remember. I never got that much better technically at playing drums, but I *did* get better at knowing what to play, when to play it, at what intensity, and with what feel. And I got much better at reading the people I was working with and how my playing could change the mood of the moment.

When people ask how I got into the music business, I tell them that story. That's how I got in. I seized a moment when I saw an opening. Honestly, I'm still not sure what "getting out there" means. And back then, social media wasn't a thing. The way I communicated with Sue had nothing to do with getting myself out there; I simply recognized she was someone I needed to connect with. She was a person who could not only send me work but also someone I could send other musicians to.

But the dreams of an 18-year-old are rarely the dreams of a 30-year-old, especially when those dreams have been realized to a reasonable extent. As I looked to create ways of making money while being with my young family (i.e., not forever on tour or falling into an underpaid teaching position), I used my knowledge of the world of sound to explore podcasting and help brilliant but introverted business owners attract clients. I began to refine a method based upon the way I built my career in one of the most competitive industries in the world. The Profitable-Pod Method has been applied by business leaders, consultants, and coaches, as well as leaders of nonprofits and even an astronaut.

I'm going to show you exactly how to use a podcast as the lever to get in there with the people you need to be talking to, in a way you're comfortable with, to grow your business, so you can make the money you deserve as an expert in your field.

THE NARROW VIEW

- Always be aware of who is in the room. Being genuinely interested in people will pay dividends for your nascent podcast. Also, being nice won't cost you a penny!
- You don't have to be in hustle mode to find an opportunity.
- If you meet someone you suspect might be a good contact for your podcast in a place where work or networking is the furthest thing from their mind, be respectful and simply ask where you can find out more about the work they do. If they are a professional, they likely have a website.
- It's always better to take details or a business card than to give yours out because it'll give you the opportunity to follow up rather than hope they call. Your position is "interested in" them as opposed to asking them to take responsibility and time for checking out your stuff.
- When you send good people to good people, they are more likely to think of you when something comes along.

1

POD POWER

Trying to disguise my panic, I gazed out the window at the drab north London weather. My business partner James and I were sitting in our undersized, overpriced serviced office with the air-con stuck on "arctic," and we were shitting ourselves.

I was recently married, with a child on the way, and the money was drying up. Having been a touring musician for over a decade, I'd just turned down the offer to be in the band for a long running musical. That show was about as much of a job-for-life as any musician could hope for; however, I didn't want to be on tour anymore. Still, I also needed to feed my family. So, this business we'd started had to work.

At first, it went well. A wedding band for hire, we got booked for gigs at least twice a month. But now, bookings were starting to slow. As great as we were at being in-demand freelance musicians, we were rookies in the world of business. The problem with the wedding market is that customer retention isn't really a thing, and referral is often tough too (newly engaged Becky Bridesmaid we met at Abi Newly-Wed's reception would tell us, "I can't book you because I need something different from Abi.") Even after James gave her the "every wedding we do is unique" speech, she wasn't buying it. We were crushed. Even the times those calls did work—all that effort for just one gig? We were used to being musicians, but managing the business side was another skill entirely.

One day, passing the time, James asked, "What's the deal with this podcast thing then?"

Since 2007, I'd been listening to podcasts on the train and in the car, especially on the way home from gigs. While my ears needed a break from my constant subconscious analysis of music, my brain needed some level of escapism, and my eyes welcomed the rest.

As I was explaining what I'd discovered so far, I had an idea. "WE need to start a podcast!"

"Why? Who for?"

"Event planners."

Within six months of launching our podcast, we were being flown around the world, playing to princes and princesses and, best of all, we were finally making real money. The best bit? We probably had under a hundred listeners. It turns out, they're not where the money is.

WHY I LOVE A POD, AND YOU SHOULD TOO

These days I advise therapists, doctors, lawyers, musicians, SaaS business owners, coaches, and business leaders who want to increase their revenue without having to grow an audience of millions. With the help of my world-class podcast production team, my clients produce podcasts that bring in leads and business opportunities, build passive income through online subscription membership, create lucrative partnership deals, retain clients, and dramatically increase revenue.

Since the pandemic of 2020, more people listen on their desktop than they did a year ago, because more people work from home now, and will likely continue doing so. At the same time, everyone and her aunt seems to have started a podcast during lockdown. Most were disasters and should never have happened. But savvy business owners started using podcasting to stay relevant and in communication with their audiences. As listeners adapt to a post-pandemic world, where and when they listen has changed, but how much they listen hasn't. In fact, podcast listenership is still growing.

Aside from the rising popularity of podcasting, another reason I love podcasts is the opportunity they present to showcase what we love. I work from my home office in a leafy suburb of London. Lockdowns aside, I'm still able to play music too. It's part of who I am, and I will always want to play. This is an important point to make because all my clients love what they do too. If they don't, it's often because the business element of work has forced them into a time-hungry management and admin hell. Podcasting does not have to stop you doing what you enjoy; it can enhance it.

We'll get into the nuts and bolts of positioning later, but consider this: your podcast doesn't even have to be directly about what you do. It may be more useful if your podcast is focused on what your suppliers do. For example, if you run a building firm, you could have a podcast about building, but it's likely you'd win more building contracts by referral from an architect. So, it might be better to have a podcast about design.

Ultimately, being famous is overrated. You only need to be well known by the people that matter. A successful podcast doesn't have to be heard by lots of people, because most "people" will never be your customers. Use your podcast as a non-threatening invitation to connect with you and shine a spotlight on the people you want to work with. After a genuine conversation, in which you prove to be interested in what your guest has to say, the way your guests listen to you and hear you will forever be changed.

> **Use your podcast as a non-threatening invitation to connect with you and shine a spotlight on the people you want to work with.**

Over the last ten years, I've discovered that, done properly, podcasts are the ultimate marketing tool for any freelance professional or business owner. Once set up, your podcast will save you time and give you the opportunity to repurpose episodes quickly and easily without huge costs. But here's the first big secret to a narrow profitable podcast: all of that "out there" marketing doesn't really matter. If you build a following, that's great, and it will no doubt feel good, but the Profitable-Pod Method regards all that stuff as a bonus.

To be perfectly clear: as a business owner, you don't need a ton of listeners to make money from a podcast. In fact, you might not need more than a couple, including you! Most people think you need to get your message out there. Well, they're wrong. The reason you should love podcasting is because it gives you and your business the opportunity to go narrow and get in there with the right people, create a meaningful connection, and find the business you want.

Finally, there's an intimacy in podcasts that other media can't achieve. I'll show you in more detail how podcasts can be versatile, like a Swiss Army knife, performing a specialized function for a specific audience. After I read Julian Treasure's book *How to Be Heard*, I started to understand why podcasts have such a profound effect on me personally. The human voice has a power like no other. This is because most of us heard our mother's voice before we saw her face. A voice without accompanying visuals often creates stronger and more emotive connections. Julian revealed to me that this wasn't a new concept; in fact, it goes back to ancient Greece. The philosopher Pythagoras was known to have put a screen between him and his students to prevent visual distractions.

WHAT'S HOLDING YOU BACK?

The fact is that many business owners are already producing content, from blogs to social media. But if you're not creating content in an audio format, you're missing a trick. And if you are producing audio but not getting the results you want, it's probably because you haven't thought it through.

From astronauts to politicians and business consultants to therapists, I've found three main issues podcasters struggle with at the beginning. But be reassured, I will address all these concerns in the book so that you will have no reason to hold back.

1. Self-Confidence

Much has been written about overcoming impostor syndrome by far more qualified people than I, so I won't cover it here. That said, if you are a genuine imposter, I can't help you! But if you've got credibility and you suffer from the "who do you think you are?" voice in your head, you'd be like every podcaster/musician/human I have ever met.

On the simplest level, many reluctant podcasters are unsure about the sound of their own voice. Will it be engaging? What about when I stutter or "umm and ahh"? Look, maybe you will sound a bit wonky, especially as you're starting out. But the great thing about a podcast is that it's not a live radio show. So, with practice and the professional editing you should be getting, all of those verbal stumbles can be taken out. And when you have guests on your podcast, you don't even need to be interesting; you just have to be interested. Come up with questions, actively listen to your guests, and you're set.

If you're nervous about your ability to address the masses, I can put your mind at ease. In the early days of radio, groups of people would gather around to listen. The same happened less than ten years later with TV broadcasts. Millions of people would be watching or listening at exactly the same time. Here's where podcasts differ. The majority of podcasts are listened to privately, and that offers you the chance to adopt a more personal speaking position: a "one to one" or "one to many ones" narrowcast, rather than the traditional "one to many" approach. That's great news, especially if you consider yourself an introvert.

You don't need to put on a big show when you speak. Your stage voice is not needed, and you do not need to be a professional performer. In fact, I spend a lot of my time reminding professional speakers to tone it down. So, if you don't identify as a confident public speaker, you may find podcasting easier than those who do! Later, we'll explore good presentation skills.

More good news for the introverts is that the Profitable-Pod Method isn't really about you at all. So, if you're concerned your podcast is a vanity project, worry not! While I have no doubt you have a lot to give and a lot to say, I understand all too well that it's way easier to do your best work with people and not with a microphone in a room alone, with only the voice inside your head for company. I'll soon be introducing you to the Pod-Lationship Cycle and how, by using it correctly, you can grow your business by putting the spotlight on others. While solo episodes (podcasts where you speak alone) are part of the method, in most cases, they should be closer to a quick voicemail than an hour-long lecture.

2. Technical Know-How

Potential podcasters also struggle with tech. They don't know what equipment they need … or what they don't need. And it's really easy to get stuck here. Many people think this is the most important part of starting a podcast, but it's not. And I'm going to make that easy for you later in the book, so you've got no more excuses.

The other technical part is how to turn audio files into a podcast. And the truth is, a podcast is not an audio file; they're quite different. For clarity, here are only three technical terms you should care about when creating any podcast:

- Preproduction: The set up and planning phase.
- Production: The recording bit.
- Postproduction: What happens after you have recorded, from editing to final publishing.

I'll address all these elements in this book.

3. Time and Energy

I often speak with potential podcasters who are worried about how much effort is needed to invest in making a great podcast. Beyond dealing with the tech, they worry it's going to be a time-suck, and they won't be able to compete with what is already out there. If that sounds like you, I have good news: none of this is important when you're narrowcasting and running the Profitable-Pod Method. So long as you're prepared to delegate the tech tasks, you'll have time to narrow your focus and get in there with the specific people who will help you grow your business, while your competitors are barking up the wrong content-creation tree.

By the end of this book, I hope you will not only feel there are no more barriers to creating your own podcast but also that it's something you simply must do. With all of these fears allayed, you should also be incredibly excited by the potential of your own podcast. So, without further ado, it's time you started thinking about YOUR podcast. Before any recording, there should be an amount of planning and prep. This is what's known as **preproduction**.

THE NARROW VIEW

- Get clear on what your ideal guests/clients want to be seen/heard talking about.
- Use your podcast to create a legitimate reason for ideal guests/clients to speak with you.
- Podcasts keep you relevant, even when you can't travel.
- You don't need a large audience for a profitable podcast. Narrow pays!
- Podcasting can save time while creating unique content.
- A narrow podcast will get you IN with the right people and create connection and business for you.
- You don't have to be an expert speaker to be a podcaster, so long as you have a good editing team in place.
- Don't stress or obsess over the tech—I'll show you what you need.

PREPRODUCTION: TAKE AIM

If you fail to plan, you are planning to fail.
— Benjamin Franklin

The preproduction phase of your podcast is primarily about what you need to do before you launch a podcast. It is also about the preparation needed before you record each episode. This "before" moment is critical to the success of your "during" (production) and "after" (postproduction) moments and will ensure that making your podcast is both profitable and pleasurable.

In this section of the book, we will explore the big preproduction questions, such as, Who is this podcast for? Why should they listen to it? What type of guests do I need? When will they listen? How can I ensure it isn't just a time-consuming vanity project? How will it help my business? You'll need to be clear on these issues before you start to actively produce audio and web assets, because your assessment of the audience will inform your entire production plan.

I will also cover how to plan each episode once you have your foundation in place, specifically how to position your show correctly before you begin and follow the simple and effective "Pod-Lationship Cycle" in order to be on your way to a profitable podcast.

2

NEVER START A PODCAST WITHOUT THESE

Children's parties get a bad rap. For most parents, they're a pain in the arse to organize, attend, or even just think about, what with the mess, cake, screaming, and everything else. But not me. I love children's parties. As a parent, any activity involving my kids is an opportunity to speak to other adults, to get out of the house/office and actually have a conversation. And conversations are one of the things I enjoy most.

At one particular kid's party, a mutual friend said, "Hey, you should meet Howard. He's got a podcast." I met Howard on a tractor ride. Turned out he didn't have a podcast at all. Howard had a bunch of audio files sitting on his computer doing nothing. He'd made a start. But like a lot of people, he was stuck there. So, I took time out of the delightful children's tractor ride to assess whether podcasting was right for him. In other words, would it improve his business and bring him new clients? Because if it wouldn't do that, it was just going to be a vanity project. You and I don't need to share a tractor ride at a children's party. But what we're going to talk about now is just as crucial for you as it was for Howard.

I'm assuming you've picked up this book because you want to use podcasting to grow your business. Or you're my mum (thanks, Mum.) But there are a few things that need to be in place for your podcast to thrive and ultimately bring you clients.

THREE PRE-PODCAST MUST-HAVES

If your podcast is ever going to fly, have all these assets at the ready:

1. A freebie your potential clients and listeners can get in exchange for their email address
2. A website to host your freebie, podcast, and products and services on
3. Clear services and/or products you sell. You are a business owner, after all …

(Caveat! I am assuming you also have the following: a decent computer that can run video conferencing calls [Zoom, etc.] without too much trouble, access to a stable high speed internet connection, and a bank account. If you don't … well, what kind of business are you running?! The point is to take some time to assess your capability before you move on.)

1. Your Freebie

Offering a freebie (also known as a lead magnet or opt-in) could take many forms. You can change the offer as often as you like, but to start I recommended going with a "guide to."

This sits on your website, and you'll mention it in all of your podcast episodes. This freebie is focused on helping your target market solve a problem they have or get a result they want. It doesn't require money, but it will cost them their email address. Those who opt in to get it should be able to consume the content quickly, say between 5 to 15 minutes.

While this permission-based marketing play is not designed to generate revenue immediately, it will generate trust and, as part of your sales cycle (I'll get to that in a moment), gives you permission to communicate and follow up with people who have identified themselves as "interested" in you. Both your podcast and lead magnet passively help interested parties to get to know you better, without you having to monitor your inbox 24/7.

Once you have an interested party's email address, it's your duty to keep in touch. Your follow-up can be as simple as letting those on your list know what's coming up on your podcast. You can also show them how to discover more about working with you.

I discourage the daily hard sell. Too many businesses disrespect those who gave them permission to be in their inbox by bombarding it with stuff. Don't be *one of those*.

2. Website

Your website will serve as a place you can be found away from the distraction of social media. It should be easy to navigate with clear links to your offers, including your podcast. Without a website, your podcast will be a vanity project costing you time and money and will not help you grow your business.

You need to easily and quickly publish your podcast in blog form on your website. If you don't already have a team or a virtual assistant to help you do this, learn how to do it, and document it so you can delegate as your business scales.

Compelling copy on your website should speak directly to your target market. It shows them the problem you can solve and/or the result you can help them get. It keeps the people you want to attract but, just as importantly, it repels those who won't be a good fit for you. In my humble opinion, many marketing issues would be solved if we spent time narrowing down and focusing on who we are not for. For example: "We help U.K. drivers find fun and safe convertibles for under £10k." This tells me this business is not interested in the American or European markets. It also tells me they work with middle-income people who don't have sky-high disposable income.

Your website navigation menu should clearly let visitors know what you've got to offer. How they first heard about you will determine where they initially click on your website. If your best client sends her best client to you, that person will just click on the contact page. In many cases, though, visitors might not first arrive on your home page. Rather, it's more likely they'll find you via a blog page you've created for a specific podcast episode.

For now, let's stick to your home page. Along with a clear benefit-driven headline, I strongly recommend you display your freebie/lead magnet above "the fold," meaning visitors can see what they can get from you without needing to scroll down the page.

WEBSITE MENU EXAMPLE

> Home – About – Podcast – Course – Consulting – Contact

Tips:

- Scrolling down the home page, content should mirror the order of the navigation menu.
- Your podcast name and most recent episodes (featuring thumbnail artwork) should appear just below the fold in gallery view.
- A button directly below the gallery view of episodes should say something along the lines of "more episodes."
- You can also display buttons that link to the major podcast platforms; for example, Listen on Apple, Spotify, Stitcher, and so on.
- Make sure that when you link to platforms away from your website, these links are set to open in a new tab so that people aren't navigating away from your website.

Your website will need to include or be connected to an automated email service or customer relationship management (CRM) system that can send email automatically after people sign up. If you have a small list, some are free (check out companies like Mailchimp, Ontraport, and Keap). If you're non-techie, it shouldn't cost you much to have someone help you set up an automated service.

Remember! Your podcast and your website should work hand in hand, so you will promote your freebie offer on your podcast and make sure both are easy to find on the website home page.

3. Clear Services and/or Products You Sell

The most important thing is an "offer that converts." If you aren't already selling a product or service, this narrowcasting method isn't for you quite yet. You need a clear offer you can deliver and, ideally, scale. And you need to know who you're making that offer to. In other words, a clearly defined target market.

For example, a dentist sells dental services to people with teeth. That market is WAY too wide for our purposes. On the other hand, if you're a dentist in Hollywood who only makes home visits to A-list celebrities to guarantee their teeth always blind the paparazzi, that's a sufficiently specific target market that can be REALLY useful for a podcast. But starting with your target market, your result and associated offer is going to save you a heap of time and, dare I say, anxiety.

> **Position your website as a pillar of your sales cycle.**

All too often, I find podcasters do not integrate their podcast properly into their website and the podcast fails in its purpose to drive business and ends up on the back burner. So, don't forget to position your website as a pillar of your sales cycle.

Simply put, a clear sales cycle shows people what you've got and how they can get it. Your sales cycle should include what people can buy but also ways they can "invest" without actually making a financial commitment. The two essential investment opportunities you should have in place on your website are the following:

1. A freebie download (as discussed). Investment = their email address.
2. Your podcast. Investment = their time spent listening to it.

Both will provide value and help you to have better sales conversations with the kind of people who can't wait to become your client. There are ways to monetize podcasts with exclusive subscription-based models. But remember, the narrowcast approach is about growing your business, not your podcast.

Finally, the Profitable-Pod Method does not suggest your podcast be a product you sell either in the form of paid subscriptions to listen or paid-for guest slots. Buying a guest slot on your podcast should only ever be offered if you can guarantee massive exposure and results. However, the method works best if your listeners and visitors to your website know that being on your podcast requires an invite. Sure, if they want to get in touch with you because they'd like to be on, that's cool; it's up to you to decide if they are a good fit. But, be warned, nine times out of ten, they won't be!

> **The narrowcast approach is about growing your business, not your podcast.**

Okay, so that's what your business needs to have in place. Now it's time to dig into your podcast plan.

PODCAST POSITIONING

Positioning is a simple concept but often misunderstood. And it's rarely used to its full potential by business owners.

Paris is positioned as the most romantic city in the world. If I'm the guy in charge of tourism for any other city on the planet, there's no amount of money I can spend on marketing to beat Paris for that reputation; lovers are going to go to Paris for their romantic break no matter what. Yes, I could try to be the most romantic destination in Australia, Guatemala, Morocco, or wherever, but I'll never beat the position that Paris already has.

On the internet, though, it's different because there are very few institutional names like Paris. This means things are wide open for you to carve your own niche and claim the top spot. Hopefully you've either already done this in your business, or at least understand the concept. However, your podcast ISN'T your business; it's just a tool of your business, a tool you can use to build connections and leverage those connections to generate revenue.

Whatever line of work you are in, I can safely claim you are in the business of creating emotions. Emotions create actions. Who you are to the people in your world will depend on the emotion you elicit from them. Take some time to think about your ideal listener. Although the Profitable-Pod Method doesn't focus on listeners to create revenue, it's important to clearly and quickly articulate the purpose of your show and who it benefits, not least because it is a massive help when it comes to getting people to accept your invite to be a guest on your podcast.

As a drummer, I had two choices if I wanted to make more money playing live shows: (1) get more gigs (i.e., work more), or (2) take better-paid gigs. A well-positioned podcast can get you in there with the right kind of people so you can make more without changing a single thing in your business or how you work. Or even better, the authority and ability to charge more so you can work less.

I did this in early 2020 with CrisisCast. As the world entered the shitstorm that was (and still is) the COVID-19 pandemic, it was clear that things were changing drastically. What wasn't clear was what exactly we could, or should, do about it. But I knew we wanted more podcast production clients, so I decided to create a podcast with the goal of getting more quality clients. The positioning was simple: for the listener, CrisisCast was the only place to learn directly from top performers what they were doing in response to the crisis. But more importantly, for the guests, CrisisCast was a way to showcase my podcast production knowledge and expertise so that if they decided to do their own podcast, my company would be the obvious choice to produce it. And you know what? Worked a treat!

So, what do you need to position your podcast?

1. Name

Keep it simple. Your name should attract the right people and repel everyone else. Don't try to be too clever here. It's more important to be specific. Most show titles I scroll past on my podcast app do not help me decide if I should listen, so it's fair to say they probably don't help other potential listeners either. A great podcast name will show the people you want listening to your show to instantly know it's for them and only them just from the title.

My go-to resource for helping clients figure out the name is my version of Michael Port's "Who and Do What" statement from *Book Yourself Solid*. It is the most complete business marketing and sales system I have ever come across. The formula goes like this:

> [SINGLE BIGGEST RESULT] for [TARGET MARKET] with [HOST NAME]

Head of Book Yourself Solid, Matthew Kimberley, has a podcast called "Marketing [RESULT] for Coaches [MARKET] with Matthew Kimberley [HOST]."

Like many other people, I have spent hours of my life scrolling through various subscription-based TV streaming platforms trying to decide what to watch during the pandemic lockdowns. "How about this one?" I'd say to my wife. She often replied, "Not sure ... doesn't look very good." Too tired to read the description or watch a trailer, this judgment was based on the cover picture and the title alone. A terrible idea, I know, but that is the reality of tired, busy parents with short attention spans. In podcast land, the picture that appears in podcatchers (the places podcasts are available to listen to) is just as important. It's called "album art."

Poorly designed podcast album artwork subliminally tells your potential listener that you don't take your business seriously. Professional podcast artwork will get the attention of passing scrollers and, more importantly, help both you and friends of the podcast be proud to do so.

Album art is not to be confused with episode art, which as the name implies is specific to each episode. While it's not imperative you create episode art, it is a must if you want to follow the Profitable-Pod Method. This will become apparent. Just be aware that only some podcatchers display episode art, while many will only display your main podcast album art. Thankfully, narrow podcasting using the Profitable-Pod Method doesn't rely on the ever-changing industry tech.

Great artwork grabs attention and creates feelings. Fonts and colors all play a part in creating a mood. If you already have brand colors and a logo, you should incorporate elements of these in your album art. If you've always hated your logo and don't feel it reflects who you are and what you do, maybe it's time for a rebrand, because your artwork can be the first point of discovery for potential guests and listeners, so it should pull focus, even if only a thumbnail image on a smartphone.

It's up to you whether you want to include your photograph in your artwork, but it can be a wise move, especially if you want to be recognized as a (and I hate this term) "thought leader" in your area.

Any production company worth its salt should help you out with podcast album and episode art.

Shameless plug alert: *Head over to https://narrowpodcasting.com if you're ready to get started.*

2. Locate the Listener

As you keep reminding yourself that your show is an "audience of many ones," think about where you'd like your listener to be in their day. The Tuesday morning dog walk, the Wednesday night workout, the Thursday morning commute, or the Friday night bedtime treat; how you speak to a close friend changes depending on if you're with them during any of those activities. In this case, your listener is your close friend. I always listen to podcasts in bed, but I won't listen to podcasts with loud sound effects or music because I want to relax and zone out. On the other hand, I have absolutely no problem listening to podcasts with music or sound effects while I'm driving or exercising. The reason your commitment to publishing once a week is so powerful is that we know podcast listeners are people of habit and will likely make time in their week to listen to your show at a certain time.

Two great websites detailing the latest podcast-listening trends are podcastinsights.com and edisonresearch.com.

3. Style and Tone

Most of the people who have a successful narrow, profitable podcast fit a similar profile. They don't necessarily want to be the star of the show but, like me, enjoy good conversations. And that's great, because you don't always have to be interestING; you can be interestED. Be wary of setting a "me, me, me" tone for your podcast!

Music is also critical to tone. Because music is so emotive, you need to use your show preproduction time to really consider how it can serve your podcast mission. Theme music, transitional music, and sound effects fall under the category of "sound design." Be as careful and attentive with your sound and your sound design as you would with your artwork. Simple and clear almost always wins the race, unless you're an experienced composer.

Other sound design elements and tools that can be used to enhance your podcast production values are:

- Music beds
- Sonic logos
- Event sounds

A music bed is a piece of music that is designed to sit behind speech. Using an effective music bed under prerecorded parts of your show, like intros and outros, will tell the listener subconsciously that they need to pay attention. They are used in everything from movie trailers and TV adverts to news bulletins and traffic updates. While your music bed can have a low pulsating beat or long notes that build slowly, it absolutely should not have a discernible melody or tune because that would distract too much from the spoken words.

On to sonic logos. You might be familiar with radio and TV adverts for computers with the Intel™ chip. This is always advertised as "Intel inside." These ads use sonic logos. Intel's sonic logo consists of four notes, each representing a syllable of "Intel inside."

Sonic logos last around two to four seconds.

Event sounds, on the other hand, are short noises that act as quick transitions between sections or when you want to highlight a point. Computer games are full of them. Bleeps, tings, dings, whooshes, bangs, cash registers … you get the idea. However, please exercise restraint. All this stuff can be as detracting and annoying as it can be helpful, and it can certainly remove that personal intimate atmosphere you're trying to create for your listener. Just like your main music choices, using a music bed and other sound design elements requires you to ask yourself the same questions about your main music choice.

Be conscious of the sound design and how it complements the tone you're using to speak to your "many ones" (listeners) in. The one-to-one tone I adopt when speaking to my young children is markedly different from the tone I adopt at the pub with my best friend. Both are genuine and authentic, but if I swapped them, neither would be appropriate. In fact, they'd be more than a little weird.

If you speak in an upbeat way to your listener and use upbeat music to match, it's likely your podcast will work great as a gym companion. That's perfect for you if you're in the business of coaching young executives how to scale their business. If, however, you teach those same young executives how to unplug and unwind with meditation, you'll be hoping your listener will adopt a more mindful and reflective state. In that case, a calmer tone and lighter music, perhaps at a slower tempo with longer notes, will help set the atmosphere.

By now you should be seeing the unique opportunity to please your audience, who will carve out that specific moment in their week just for you. This is the best result you can get when you think about listeners, because the problem with podcasts is you don't know who these listeners actually are unless you have a way of getting them into your life. This is why a clear call to action on your podcast to your website where people can download a free guide is so important.

Whatever you do, do not leave your soundtrack in the hands of a cheap do-it-all podcast professional and trust them with your brand. Be involved in every element of your podcast branding and positioning, music included. Not all professional production companies have experienced musicians on staff who make sure music is matched well to your show … and that it's being used legally! I do know of one though: https://narrowpodcasting.com.

THE NARROW VIEW

- Have a clear free offer.
- Have a website that quickly shows anyone the result you offer your customers.
- Make sure your website has the facility to create a blog entry for each podcast episode.
- Put yourself in the space of your listener. Think of how they'd want to be spoken to. What sort of content will they enjoy?
- Nail your podcast name before all else, and use it to inform your artwork and soundscape.

3

MIND YOUR BUSINESS (MAKING PODCASTING WORK)

A band is made up of a drummer and a bassist with additional novelty acts.
—Nick Mason, drummer, Pink Floyd

Drummers often look around and see people who are wealthier, more successful and famous, but with half our skill or expertise. They often get really narked about that, which is totally understandable (I'm not immune to that myself). It's true in virtually every sector of society that those who are the best self-promoters get massive results, while those who are actually laying the foundation of a business (or band) are often not recognized for their skill.

Hopefully by now you've decided that podcasting *is* right for you, and with the Profitable-Pod Method, you can be the drummer skillfully laying down the foundation of groove, subtly forging your own distinct sound without having to get into the sleazy salesperson tactics of a lead singer or guitarist. So, let's get into it.

A lot of business coaches talk about vision, mission, brand values, and a whole bunch of other nebulous concepts. And for a lot of beginner podcasters, that can sound really appealing. After all, it allows them to procrastinate for ages! But even if they do eventually launch, it plays into the fact that they love the sound of their own voice. Hopefully, at this point in my book, we can be clear that's not you; you want to actually make a profit for your business using your podcast.

Making a profit is certainly possible, so let's talk about the three types of podcaster who succeed in making money from the show:

1. You're already famous and want to boost your profile.

Whenever Michelle Obama or whoever hits the *New York Times* bestseller list decides to launch a podcast, it's simple. They will immediately pick up sponsorship from at least one, maybe more big companies. And that ensures an income stream right off the bat.

2. You're a creative who is looking to get paid for your passion.

There are people who develop their podcasts over a long period of time, and it's a purely creative endeavor for them. Most never make it into the big time, but one or two will get picked up by networks like iHeart and even Netflix or Amazon who look to podcasts as a proving ground for new talent. And the lucky and talented have their ideas turned into a TV show.

This is kind of like any big artist or band. Before they got big, they likely played A LOT of pub gigs. Each week, they would try new material with a different audience and measure the response. Then over time, their "content" is refined. And though the vast majority never make it, one or two do and get signed to the big labels.

3. You're a caring business owner looking to scale.

At this point, you've probably realized that neither of the previous two paths is for you. And believe me, that's a good thing. If I had a pound coin for every time somebody told me they wanted to be the next Joe Rogan, I'd have a private jet on standby.

The caring business owner isn't interested in fame, or even fortune, really. They just want to grow their business, serving their target market the best way they know how, and enjoy what they do along the way.

These other two paths require you to get OUT there. The final path—the Profitable-Pod Method—is getting IN there with your ideal clients, and the way to generate income is giving your guests a good listening to and then leveraging those relationships to grow your business.

> **The way to generate income is giving your guests a good listening to and then leveraging those relationships to grow your business.**

GOOD, BETTER, BEST

Who should you be building relationships with? And how do you use a narrow podcast to do that?

Most beginner podcasters are vain, wanting to be famous and get out there. These people don't even consider the quality of the guests for their show because it's all about them and the sound of their voice, rather than their listeners or, crucially, their business.

The other common problem is people who are just so grateful that anybody would even talk to them that they lose the point of why they are podcasting at all. They bask in a warm glow of gratitude for the fact that somebody, anybody will be a guest. Neither of these people will make any money for their business from their podcast.

What I'm interested in is the "good, better, and best" outcome that having somebody on your show can achieve. That way, we're always tactically thinking about who to invite and what value we can co-create with them.

To approach potential guests, you have to know in advance why they'd say yes to you. The Pod-Lationship Cycle covered later will help you do that, but for starters, let's be clear on the only three reasons to invite a guest on your podcast.

1. They're an existing client of or prospect for you/your business.
2. They're a powerful referrer, either for somebody you specifically want to be referred to or because they have a network you'd love to have access to.
3. They're a potential joint venture partner for you.

You should absolutely NOT be talking to anyone else. Stay narrow! And if you get only this bit right, you'll automatically be on the way to a profitable pod because **the "best" outcome** is that they will hire you, refer you, or partner with you. And that's likely best for both you AND them. Therefore, it's a win-win.

The "better" outcome is that after the podcast is published, they actively talk to people in your target market about you. They're already connected to your target market and regularly having conversations with them. Now they're talking about you, which is, of course, a major objective for any expert running a small business.

The "good" outcome is that they share the episode with people. For example, they share the episode they guested on with art and a link to your blog with words like, "I was recently featured on this podcast. It was a great conversation, check it out." If they do that, then you're winning already.

If you can't see any of these outcomes being likely from someone you're thinking of inviting onto your podcast, don't invite the person. Keep that in mind, and you'll be running a profitable pod in no time.

ARE YOU A ONE-MAN BAND?

There tends to be two kinds of entrepreneurs who start a new podcast. The first kind are the people making between six and seven figures consistently who generally tend to have at least one virtual assistant (VA). Or maybe they regularly outsource things they don't love doing. In my experience, these people have figured out their offer and it's converting regularly. But they recognize other experts and are quick to invest in things that allow them to scale WITHOUT doing any more work.

On the other hand, there are the people making less than six figures who aren't in the habit of outsourcing stuff. They probably don't have a team and tend to do everything in their business. In my experience, that's usually because they haven't fully dialed into their target market and/or their offer. These people tend to want to save money by investing time and doing everything themselves. For a single, hour-long audio file, you need around 12 hours to produce a finished podcast episode. That includes the following:

- Editing and sound production with good levels, and so on
- Good search engine optimization (SEO) based show notes (more on those later) that make you more visible on the internet
- Creating episode artwork
- Publishing the episode on hosting platforms
- Updating the website to include the episode blog
- Repurposing content into shareable social media assets

Now, obviously as a partner in the world's best podcast production company, I'm going to recommend that—for the love of God—pay somebody else to do all this for you. That way you can simply focus on doing what you do best and allow your narrowcast podcast to cultivate and grow the relationships we've talked about. But on the off chance you really think doing it yourself is the way to go, now you know what's involved!

I've told my team to look very kindly on readers of this book who follow this link: https://narrowpodcasting.com.

THE NARROW VIEW

- Don't invite guests who aren't in or around your ideal clients. Just because you can speak to them doesn't mean you should!
- Get clear on the "good, better, best" outcomes of speaking with your guests.
- If it's financially viable, hire a professional production company to work on your podcast. As a business owner, your time is better spent elsewhere.

4

CUTTING THROUGH THE NOISE

It might seem obvious, but noise equals unwanted sound. As a drummer, I know a fair amount about noise. When you're trying to play music, unwanted noise is the enemy. You may have seen the plexiglass screens that they sometimes put around drummers on stages and in orchestra pits. The reason they do that is to minimize unwanted sound from getting into microphones where it doesn't belong. When I set up next to a bass guitar player who sings, sound can get messy quickly. Sound from her amplifier can go down into my drum microphones and cause problems with the drum sound, or the sound from my drums can enter the microphone she sings into and drown out her voice. And all of that is before we even consider the noise of an audience.

Anything getting into a microphone that isn't music makes it harder for the music to be heard. We'll talk more about how to deal with the tech aspect of this in the next chapter. But before that, I need to touch upon the other kind of noise you're already acutely aware of.

The amount of "noise" out there on the internet is at an all-time high. Between social media, push notifications, and email, the average person touches their phone around 3000 times a day. And it's getting louder by the day. Podcasts are part of the cacophony, and they multiply by the day. And, just like with drumming, being the loudest isn't the answer. Instead, you must know how to cut through the noise completely.

There are five ways you can stand a better chance of being heard.

1. KEEP IT SHORT 'N' SWEET

I'm often asked how long a podcast should be, and my answer sometimes upsets the brilliant-but-unknown people who ask. Most podcasts are too long. They are self-indulgent audio vanity mirrors that do nothing for anyone.

The sad fact is, most podcasters forget that unless they already have a bestselling book or they're already well known, prospective listeners (and clients) will not invest lots of time in checking out their show. But what about Joe Rogan and Tim Ferriss? I hear you ask. Yes. These are great podcasts, and many episodes are over two hours long. Hosts like Joe Rogan and Marc Maron started in 2009 when the podcast landscape was much smaller, and Tim Ferriss had a *New York Times* bestselling book before he started his podcast. You're not Joe Rogan, Tim Ferriss, or Marc Maron. But fear not. Compared to other media like YouTube, it's still early and you can still make waves on a beach you own.

When it comes to podcast length, remember that listeners likely won't invest much time in you early on in a relationship. So, what's the answer for people like us? Publish short podcasts. They're really valuable.

Respect listeners' time.

It was the author of *Book Yourself Solid,* Michael Port, who first taught me that the amount of time people will invest in you is proportionate to the amount of trust you've earned. A short solo episode positioning you as the expert is both cool and way easier to make than a longer one. It's better to ask potential new listeners to invest only a "10-minute dog walk" in you or a "12-minute exercise bike" or a "15-minute drive to pick the kids up." These short podcasts are perfect for most people's daily journeys or chores. They are much less likely to listen to these solo episodes, especially early in your podcasting life, if they are longer. In short, respect listeners' time. For a while, the author and founder of CD Baby, Derek Sivers, made two-minute-long podcasts. Seek them out for inspiration on how to be concise.

With your interview shows, 30 to 45 minutes is more than enough for a final edit. This often means that the raw audio from your original conversation is more like an hour long. Once you (or preferably your production team) have edited out the verbal clutter, you'll get something that is super valuable. If you manage to secure a guest who you know will bring listeners with them, longer interviews are fine. If they are really well known, you may even manage to squeeze two episodes out of one long conversation.

That's the "short" part covered, but when it comes to the "sweet," it's about being useful. Marc Maron interviews people I know I will find interesting, so I listen when I'm in bed. And guess what? I almost always fall asleep during the interview because each episode is so long. I don't always go back to the episode because, as much as I love his podcast, it's not essential listening for me. I say all this because I want to make sure that you are at least aware of the differences between interesting and useful.

If your podcast is framed as useful to your target market, or an adjacent one, people will most likely listen during the day rather than use it to help them nod off at night. I listen to self-development podcasts when I ride my bike for around 40 minutes most days. I'll listen to podcasts that I think are going to develop me and my ability to help my clients. I also listen to self-development style podcasts when I travel to play a gig. On the way back from said gig, I'll likely listen to comedy or fiction to stay awake because, by then, I'm rarely in the mood to focus on strategy.

> I want to make sure that you are at least aware of the differences between interesting and useful.

2. EYE-CATCHING TITLES AND GUESTS

Remember that people won't listen to your podcast like they watch a Netflix season. Because they can join any time, you should account for that by ensuring each episode has a title that entices your people to give your podcast a try. That often includes mentioning the guest.

For example: Improve Your Sourdough Bread Skills with Gina Lou Tone.

When you have a guest, always include their name in the episode title. Simply take your headline and add "with Guest Name." It helps with search results online and puts the spotlight on the person that matters.

Just like your business, your podcast show should be labeled clearly with a great name showing people exactly what is inside your podcast tin. Beyond that, you should name your podcast episodes to further explain the benefits of listening. Think of each episode name as a headline. There are various types of headlines you can use to attract your listener.

- "Toward" or "Gain" headlines will often start with "How to…" or "Discover."
- "Away from" or "Threat" headlines are designed to induce fear of missing out (FOMO) and can start with prefixes such as "Do you recognize," "Why you shouldn't," or even "Do not."

You can also use social proof/evidence in your headlines. For example, if I was a guest on your podcast, an appropriate title could be "Discover how Toby Goodman helps his clients cut 99% of marketing activity and still grow their business."

3. LAUNCH STRATEGY: THE '10' GAMBIT

When you launch with ten episodes up front, you give your listener ten reasons to check your new show. This buffet approach should ideally split into five short "authority episodes." These are solo shows where you get to demonstrate your credibility without asking for too much time from your new listener. They are also highly shareable to your existing network and can act as a great client reengagement or retention strategy.

Your other five shows should be guest interviews. These episodes can be longer because your guest will share them with people who are interested in them.

All the shows should be well produced. Ideally, you should be using a production company who understands the Profitable-Pod Method to produce your show. So far, I only know of one … narrowpodcasting.com.

4. CREATE A TRAILER

There are four reasons you should create a one-minute trailer for your podcast.

1. It forces you to be clear on what your podcast is about and who it is for.
2. It ensures that when you're ready to start publishing, all the platforms have accepted your show because your RSS is complete, it syndicates to all the right places, and ensures distribution to all platforms automatically.
3. You can use it to let your network know what's coming.
4. It is a great, professional-looking tool you can use to invite your ideal guests onto the show.

5. GET PERSONAL

Being in someone's ears in the "me time" moment of their day is worth it. Podcasts are better than visual media in almost any setting (gym, car, walking the dog, etc.), for being personal to the listener. They become the fly on the wall, privy to a private conversation between two people. It's natural to slip into presentation mode, but avoid the "Hi everyone" trap because it'll never be as personal and connect as powerfully as "Welcome, thanks for joining me."

From an 'out there' marketing standpoint, the key is to plan your episodes with a single objective: Being in the ear of your listener, for 10 to 30 minutes, every single week. The nature of podcasts means you get to spend valuable one-to-one time with your listeners regularly without having to record every week because you can batch-record several episodes ahead of time. And that keeps you at the top of the podcast feed on your listener's device and in their life on a regular basis.

> **Podcasts are better than visual media in almost any setting, for being personal to the listener.**

THE NARROW VIEW

- Create a trailer for your show that conveys the content and the tone/style accurately.
- Record an intro and outro. You'll use it on every episode!
- Launch with ten short 'n' sweet episodes ready to go. That's five interviews at 30–45 minutes each, edited down from approximately 1 hour, and five short solo episodes of between 5–15 minutes. These position you as an expert and act like a buffet people can choose from to get to know you.
- Give your episodes great titles, and book appealing guests.
- Plan to publish an episode weekly: Alternate guest interviews and solo episodes work well.
- Create an introduction to your free guide (lead magnet), giving your ideal listeners a reason to visit your website.
- Design a concise landing page to house that lead magnet enabling listeners to identify themselves to you so that you can deepen your relationship with them.
- Above all, respect your listeners' time.

5

FIVE LITTLE POD DUCKS

I realize it is a little trite to tell you to get your ducks in a row before you launch, but I'm going to anyway. Here in leafy Hertfordshire, England, there are a ton of ducks. I often see them while I'm out cycling around listening to podcasts … and, yes, they're always in a row. But the thing people often forget is that the baby ducks are in a row BEHIND the big mummy duck. They're following. With your little pod ducks, however, they are all equal. They are still in a row but walking side by side. If you get them in line, your big, fat, profitable pod can follow in their wake and effortlessly help you get your next client.

DUCK ONE: BUILD A HOME FOR YOUR PODCAST ONLINE

We've already addressed the importance of your website in running a profitable podcast. But let's look at how to piece all of this together.

- Your podcast needs to sit on a blog on your business website (the podcast doesn't need its own). By being on your main business site, it builds credibility and shows you're not just somebody who has a podcast on one of the popular platforms.
- You own your website—all the content on it. You have all the passwords. You have control! Sometimes people will visit your site because they've heard you on the podcast. Other times, they might make contact with you about something else. But because you have a website, you can send them somewhere specific to listen to an episode about something that might be of use to them. And that deepens that sense of "Okay, this person actually knows what they're talking about."
- If you recall from earlier, I told you that you needed to feature your podcast on your home page as well as whatever separate page you're going to create. You also know your home page needs to contain your freebie. Adding your bio to the home page is fine, too, and testimonial quotes are even better. But here's the important part. You should structure the URL (your specific website address and page) like this: **yourwebsite.com/podcast**
- For individual episodes, create a new blog section/index in your website. Name the blog "podcast" and use the name of each episode for each blog to it so each entry looks like this: **yourwebsite.com/podcast/name-of-show-with-guest-name.**
- From a Profitable-Pod Method standpoint, it's important because it'll help you look professional to the guest. By including their name in the URL, you have another chance to impress your guest and increase the chance of getting found by people searching for them on the internet. I'll get to exactly how you tell your guest about the episode later on.
- Make sure also that you feature the podcast player at the top of the blog post so that your website visitors can see it's more than a blog about a podcast. The podcast itself should be playable from that page too.

DUCK TWO: CREATE YOUR PIPELINE

The pipeline is the entire episode creation process, in task order, of creating each episode. Systematizing and scheduling it will make your life much easier. The process includes the following:

Planning/Pre-Production

- Research (the topic and potential guests)
- Writing (creating a script for solo shows or a list of questions for an interview)
- Scheduling (the guest)

Making/Production

- Recording raw material
- Editing the episode and creating an episode trailer
- Creating the episode artwork

Distributing/Publishing

- Publishing the episode
- Promoting the episode

Without a pipeline, you're just winging it and won't get the outcome you want. The most profitable narrow podcasts I know are published weekly. I don't recommend seasons, but I do recommend batch recording. One month ahead is good; six weeks is better. And you want to have a way of organizing it all. I use Trello, an online list-making application that allows you to create project boards. Believe me, such applications (and others are out there) are indispensable organizing tools for what is otherwise a highly complex process. When you start recording a month ahead, you give yourself time to make the best episodes possible.

> **The whole point is for the guest to have a great time.**

The goal is to give the guest the best airtime possible. Make time, therefore, to do plenty of research. The more research you do on your guest and the topic at hand, the better time the guest will have. And remember, the whole point is for the guest to have a great time.

The other key advantage to running everything through a pipeline is the space it affords you. Nobody wants to be chained to their business 24/7. And when you're following a process from invite, to booking, to research, to recording, to writing and asset creation, to follow up, you will save enough time to focus on becoming one of the top people in your market.

DUCK THREE: TECH BASICS

We all know that eyes get tired. But many of us forget that ears get tired too.

The explosion in video calls over the last couple of years has drastically lowered the bar. Many people think it's acceptable to talk to you via a cheap pair of headphones with a small in-line mic. That is okay for a five-minute check-in with a client or colleague, but not for professionals who are paid a lot for what they have to say. Like me, they spend AT LEAST a couple of hours most days on video calls. And when you remember that people's ears get tired, you quickly realize the importance of good sound.

Now some people don't like having things in their ears. If that's you, fair enough. And although Bluetooth, AirPods, and the like are good, they can never create the full rich sound of a wired broadcast-quality microphone. A nice angle poise mic arm will help, too, because it'll keep your desk clear, and you can angle your microphone so you don't transmit that bloody awful sound of typing that obscures speech. Without the arm to minimize desk noise, there's almost no point in having all the other stuff.

For a more NPR or BBC style sound, getting a mic with an XLR output, that is, a "proper mic," is worth the extra money, but it also means buying an "audio interface." An audio interface is a small electronic box that converts the signal from your microphone into your computer. I use a Focusrite Scarlett 2i2, USB interface for audio. It's not mega money, but it changes how I'm heard. And that is, of course, the entire point.

I also use headphones when I record. Using them means the only audible sound in the room is meant for the microphone, and I don't get a feedback loop from my computer speakers. Also, because the headphones send some of my own voice back into my ears, I automatically speak quietly. This not only saves you from voice strain, it is also easier on the listeners' ears.

Headphones, therefore, force you to talk in a softer, more personal tone, but one that can still be heard clearly. *And because a lot of people listen to podcasts on headphones, and you're talking directly into someone's ear, you are, effectively, a voice inside their head.* Get that right, and it's powerful stuff.

> **You are, effectively, a voice inside their head.**

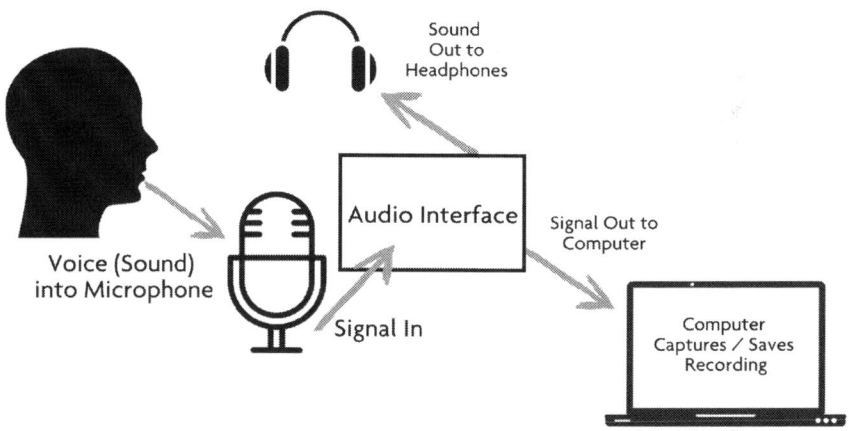

Fig. 1: Sound process with audio interface

Lastly, the room you record in makes all the difference. You could have a million dollars' worth of gear, but a bad room will make your podcast sound awful. The room you're in needs to be as silent as possible. Close windows, turn machines off, and keep your device and phone notifications off (including vibrate). Record in a room where there is no echo—blankets and soft furnishings help, but there are professional products you can buy to reduce echo. During the pandemic, the host of the popular podcast *This American Life*, Ira Glass, shared a picture of himself recording in his closet. You can do the same while you're waiting for something more permanent and comfy!

DUCK FOUR: LEGAL BASICS

> **You don't own anything on your podcast unless you have your own RSS feed.**

As a business owner, you no doubt understand the importance of rights and intellectual property ownership. What's important to know about podcasting is that you don't own anything on your podcast unless you have your own RSS feed. You can only get one by having an account with a decent hosting company. Back in chapter 2, I told you what you must have in place before you started your podcast. A subscription to a podcast hosting company is what you must have to start and maintain your podcast.

The second issue to be aware of is podcast networks. If you're invited to join a network, especially early on in your journey, beware. At best, they come with risk; at worst, they're scams. Podcast networks house a group of podcasts, which can either be credible or nefarious. Credible networks like Radiotopia and Double Elvis all host shows with related themes. Nefarious networks are commonly set up to profit the network owner only and will involve hijacking content (so you don't own your podcast) and manipulating your podcast audio with additional adverts promoting the network rather than your show or your business. Either way, unless you're careful, it is likely you won't actually own your podcast if you join a network. If you're looking for help on hosting, head over to narrowpodcasting.com, where I can hook you up with the best in class.

I'm sometimes asked about what happens when a guest wants to retain the rights to something they said. From a narrowcasting point of view, these people are not the guests you want. In practice, you're asking your guest to trust you. And when they agree to be on your show, you can ensure they understand that they are agreeing to let you edit the material they provide. If they don't know you, it's your responsibility to reassure them that you won't manipulate the editing to make them sound a certain way. I would hope that was obvious, but it's worth stating! You're looking for a relationship, not an exposé, so never create your own narrative in the edit or remove vital context from your guest's words.

> **Don't be duped into opting for free podcast hosting.**

Certain guests might need to talk to somebody in their legal department. In this case, typically we're talking about employees at big corporations, but there might be other people who have to run something past lawyers. If that's the case, you might want to use a release form. This is when I always recommend the excellent Gordon Firemark. He's a lawyer in the podcasting space who provides a release form template on his website. If you're looking for a more in-depth background of the legal issues around podcasting, he also has a ton more information at podcastrelease.com.

However, Todd Cochrane, founder of Blubrry, a hosting platform whose technology has helped over 100,000 podcasters publish their shows, has proudly never signed a release form. I'm with Todd on this one because if you're interviewing people on your podcast who demand to review your content, you are either asking the wrong way or you're asking the wrong person.

Finally on the topic of legal basics, YOU MUST make sure that all music and artwork is either original, rights-free, or used with permission.

DUCK FIVE: INTRO/OUTRO

Podcast assets are vital to the success of any pod, especially a profitable one. In terms of the prerecorded assets, we're talking about an intro and outro. The most effective way to communicate with listeners and potential guests is to frame your show in a way that is quick and easy to understand.

You can use the following scripts to model your intro and outro. Intro:

> "**Welcome to** [show name], **for** [target market] **who wants** [single biggest result]. **I'm** [name] **from** [mywebsite.com], **and in this show we'll explore** [benefit], [benefit], **and** [benefit]."

(Your freebie / lead magnet script should come after your intro, but this is not a prerecorded asset because you'll likely change it on a semi-regular basis.)
Outro:

> "**Thanks for listening to** [show name]. **The podcast can be found everywhere podcasts are available. You can check out all the links and resources mentioned and catch up on all episodes at** [mywebsite.com]."

By doing this, you're bookending your podcast with your website, which gives the listener two opportunities to follow up. It takes advantage of the rule of primacy and recency. As a podcaster, you want your listeners to remember what they heard at the beginning and what they heard at the end. And if it's the same thing, that will really help, because sometimes your listeners can be distracted mid podcast. If they are listening on their phone, it may go off and take them away from you. If the only action your listeners can take to discover more is to go to your website, they are way more likely to remember it and take the action. If your website is easy to say and easy to spell, all the better.

Record your intro and outro and have them mixed so your voice sounds amazing. Maybe you change them once a year, or maybe never, but remember that the primary task of the intro/outro is to embed your URL in the listeners' memory.

Adding music to the intro/outro is a choice. There are a growing number of places where you can get license-free music. (Do not even think about using music without express permission.) If you're looking for music at a bargain price, there are a number of stock music websites you can use. However, with stock library music, you always risk hearing "your music" on someone else's podcast, radio ad, and so on. If that isn't an option for you, it's well worth paying a composer to create something exclusively for you. But when it comes to music choice, tread carefully if you're going it alone. Take some time to think about the mood you want your listeners to get into when they hear your music. Of course, the choice of music should match the tone/style that you have planned for your podcast and reflect where you expect your listener to be (in bed, on a treadmill, in the car …). This is all fairly subjective, but I'd urge you to consider the kind of music you would choose to help get yourself in a certain mood. After all, we often work best with those whose interests and tastes are similar to ours. Or, as entrepreneur, author, and pioneer of commercial email, Seth Godin once said, "People like us, do things like this."

Over at my production company, part of our onboarding process involves our inhouse composers creating custom music or, where budget is restricted, they will choose royalty-free music to match the podcaster's vocal tone.

That's it! All the ducks are good to go! Now you're ready to get producing.

THE NARROW VIEW

- Get your online pod hub ready with a podcast page, and then post your freebie offer on the home page and menu. Bookend your intro/outro with a mention of the website.
- Get a podcast hosting account.
- Lay down your pipeline process clearly, and plan your recordings at least one month ahead.
- Buy the right recording kit, and find the right recording room.
- Create original podcast assets that you own.

PRODUCTION: HITTING THE TARGET

If preproduction is the "before" and postproduction is the "after," production itself is the "during" moment … recording the podcast episode. Everything you've done so far has led up to the production moment.

Your performance and that of your guests will become the raw material you use to create your published podcast. Get it right and your podcast has the capacity to live on for years, paying you beyond what you ever imagined.

While you'll have the ability to re-record solo shows until you get the right take, it's less likely you'll get that opportunity with your guests, so it's critical you are in control of this process: it's the bit you cannot delegate if you really want to create a podcast that generates revenue for your business.

6

MIC DROP: PRESENT LIKE A PRO

In an old north London neighborhood, not far from where I live, there's a Romanesque building that was designed by the same Victorian architect who envisioned the Natural History Museum. It was originally a church and missionary school but in the early 1990s became AIR Studios. It's one of the few places left in the world with full orchestral recording capabilities. Many of the most iconic movie soundtracks of the past 20 years were recorded there. As you might imagine, it's got a huge reputation. On one occasion, I was working with a singer there. She was recording a ballad … and totally ruining it. It was gutting to watch. But the engineer knew the reputation of the iconic building was getting to the artist, and she knew exactly what to do. She turned off all the lights in the studio, plunging the singer into total darkness. Without any distractions, the singer could give a very personal performance. And she did; she nailed it.

Now you have all your ducks in a row, you're ready to start recording. There are

The key is knowing how to set the conditions for a brilliant performance.

several ways you can create the optimal conditions for the best performance possible. Using the Profitable-Pod Method, both solo and guest episodes will help you grow your business. Whichever type of show you are producing, it's vital to frame your episode.

EPISODE SCRIPT

You should already have your intro and outro scripts (as mentioned in the previous chapter). Now, so that your listener is hooked, you need an opening for the episode that is effective. Each episode will have some written/planned script, even if it's a guest interview. Broadly speaking, the script will include the following elements:

Part 1: Freebie/Lead Magnet Plug

This comes after your intro, and you might make it a prerecorded asset (you can change this as often as you like, although I wouldn't recommended more frequently than every 90 days, especially if you don't have a team).

As noted earlier, a lead magnet is a way to provide valuable information to a prospect in exchange for their email address, thus giving you permission to communicate with them in the future. With lead magnets, people won't invest a lot of time in you early in the relationship, so you don't need to make them complicated. The aim is an email address, not a sale at this point.

By recording your lead magnet promotion in your podcast as a separate piece of audio, you or (preferably) your podcast team won't have to worry about rebuilding your entire intro/outro every time you change the content of your lead magnet. (Remember, the lead magnet itself will need to live on your website where visitors can sign up to get it.)

Your lead magnet could be a short document, a short video, or even a series of automated messages. For example:

> **FIVE WAYS TO OVERCOME NEW-JOB ANXIETY IN FIVE DAYS**
> Over the next five days, get a tip a day to keep you on top of your first week at your new job.

While cold lead generation via email isn't the focus of the Profitable-Pod Method, it would be remiss of me not to help you build your email list. As your business grows, you'll find your list becomes a much more important part of how you make money. A list of prospects is a valuable asset in itself, but think of your lead magnet as planting a seed for the future. If you're looking at selling your business at any point down the line, your potential buyers will see value and potential value in a list of people who have identified themselves to you (and your potential investors) as people with a specific problem that can be solved or interest that can be met.

With your lead magnet, you're simply proving you know what you're talking about and that you can be trusted. An easy, fast way of writing your lead magnet is to think about a few quick tips you'd give someone you met in passing who had asked for some wise words. Consider this scenario:

> Ava runs a recruitment business helping women get jobs in agricultural farming. Ava has only two minutes to help Raya, who is patiently waiting to meet her after a conference speech. In those two minutes, Ava can give five quick tips to help Raya *start* to move toward getting her next job. The key word here is "start." Ava can only help get Raya started in the few moments she has. If she gives a few good pointers, chances are Raya will sign up to Ava's recruitment agency. The frustrating part is Ava has a flight to catch and doesn't have any time to speak to the seven others waiting in line. If only she had a freebie offer on her site!

If Ava helps women find jobs in agriculture, a jobs section should be a part of her website. Farmers who want a stronger female workforce pay Ava to advertise, and women seeking to start or develop their careers in farming pay a subscription to get access to the jobs board.

While Ava's podcast can speak to both of these people, her lead magnet (that can be changed any time) should probably only speak to one of her target audiences. For example, if she needs more farmers to use her jobs board, she might have a lead magnet, delivered as a short download, called:

How to Narrow the Gender Gap and Attract the Best Female Agricultural Talent to Your Farm. Whereas, if Ava wants more traffic from the women like Raya looking for jobs in the sector, her lead magnet would need to appeal to the jobseeker.

You need a script that will last 30 or so seconds, positioning your freebie offer just after your official intro and before your unique opener. Example script for lead magnet:

> "Before we get started, I have a gift for you. As a woman in agriculture, you understand all too well how under-represented we are and how challenging it can be to be considered for new roles. So, I've put together a short guide just for you. It's called, **3 Ways to Make Your Resumé Stand Out in the Field** and it'll help you get started on the road to finding your dream job. You can get it right now, over at [mywebsite.com]."

Part 2: Opener

This will come just after your freebie/lead magnet plug. Here's an example where brand design professionals are the target:

> "In this show, I speak with Jimimah Smith, who shares a secret all brand design professionals should know, guaranteed to cut sales time in half compared with your competitors. She also discusses how learning to swim changed her entire business model, and the three-point system she uses to create loyalty from freelancers. Enjoy my conversation with the fantastic Jimimah Smith."

So that's your opener. Let's break it down.

    a) It opens (and closes) with the name of the guest. This puts them at center stage, not you, and this will encourage your guest to share

the podcast from the get-go. People like us care way more about this than the number of listens.
b) It creates intrigue around a specific result and elicits an emotion when it mentions the competitor.
c) It creates a "pattern interrupt" that makes people super curious. What on earth has swimming got to do with brand design? By throwing a curveball, your opener helps change your listener's natural pattern of thought. When you get it right, it'll keep them listening.
d) It excites the listener with a hint about a three-point system that solves a specific problem, which suggests it will be a value-packed episode.

On a guest interview show, you should ideally record your opener soon after you've done the interview. Ask yourself what the five key points of the conversation were, and you'll be able to write it very quickly. If you leave it too long, you probably won't remember.

And as much as you like the idea, having to listen to the whole thing again is something you're likely to be too busy to do! If you have a producer on the interview with you running the tech aspects of the recording, ask them what they took away from the podcast. Hell, why not ask your guest just after the interview has officially stopped? The following question keeps things positive: "What do you feel listeners will find most useful about our conversation?"

Part 3: Script Body

If you're doing a solo show, you might choose to write out your entire script and record it verbatim, or you might semi-freestyle using only bullet points. If you're reading a script, practice it so that it doesn't sound flat. As you get more experienced, bullet points allow you to be more flexible in your delivery. Don't worry if you fluff it; re-record and let the editor smooth it over! Whatever you do, don't try winging it solo. It's likely to be all over the place.

With an interview show, you of course need your list of questions! But be prepared for tangents, since conversations are rarely straightforward Q&As, but try to bring the guest back to the important points. Don't be afraid to return to those points during the interview to ensure you've completed your questioning.

On the subject of interviewing, remember that you won't always know what they know. But, of course, they won't know everything you know. Therefore, the key thing to remember is to position yourself as a partner. They might be the expert, but so are you, just at different things. If you come on too strong as a sycophant fanboy/girl, you're likely to ruin that positioning. And that's important for two reasons: you want to elicit the best possible performance from them, but you also want to set things up nicely for your post-coital pod moment ... which we'll get to shortly.

Part 4: Conclusion/Closer

Your closer plays before your official outro. You may not need it, but you can give reminders of some of the key points, tell the listener where they can find out more about the guest, or say thank you to the guest and anyone else who made the interview possible.

TONE AND EMOTION SCRIPT

If you're reading this book, you're in the business of creating feelings, because that's all any of your prospects or customers actually want. Every action they take is motivated by a feeling. Conversely, every feeling you create during your communication has the opportunity to lead to an action that could well lead to a positive result for you. They're listening to you because they've decided either it will make them feel something, or it will help them get closer to a result they want.

Film director Alfred Hitchcock was one of the first to realize the power of an emotion script that ran parallel to the shooting script that tracked how he wanted the audience to feel at each point in the movie. In Mr. Hitchcock's case, these emotions were often fear, anxiety, stress, tension, but also hope, relief, and back again.

This is an idea podcasters can use. When you're asking people a question, or (ideally) if you've planned those questions in advance, think about the feelings not only of the listener but the guest too. What (if anything) do you want them to emote as they are talking? And, in turn, what emotions do you want the audience to pick up? Do you want them to sympathize or empathize with the guest? What actions will be taken by your guest and your listeners as a result of the feelings you create? This requires the interviewee to be open with you in the interview, but also in the period after the recording.

Using great questions and emotive music, you'll help guests and listeners feel they're in safe and capable hands. When you connect with the right guest or listener in the right moment, you'll give them a desire to take action and a desire to find support while taking it. Which is where you come in. The more time you take to work out what feelings you need to create to get people to take the actions you desire, the more effective you will be. However, always use your power responsibly and respectfully.

SOUND QUALITY: THE TECHIE BIT

One of the most important ways to control your tone is technical. You want to make sure you maintain the right distance from your mic, and that's typically about an inch away. Too far, or even too close, and you'll ruin the sound.

Either you have an "end address" microphone or a "side address" microphone. An end address mic is the kind a singer performs on stage with. The mic is pointed directly at them. Side address mics are used in studio settings. Firstly, know what you have, and then make sure to use it properly. Beware—what you think might be a side address mic could be an end address and vice-versa.

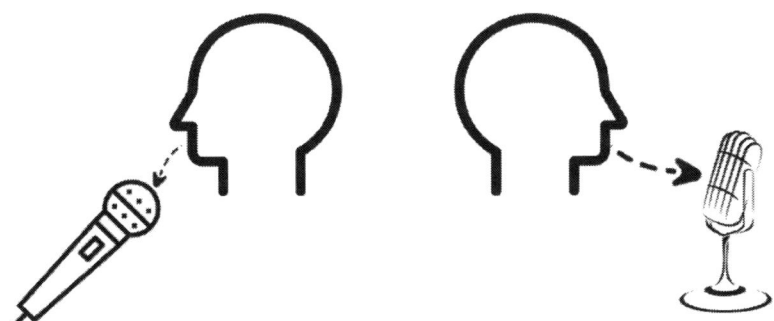

End Address Microphone   Side Address Microphone

Fig. 2: Two types of mic

Next, think about the "pop shield." That's a circle of stretched material that goes in front of your mic, which is there to stop plosives. A plosive is a sudden release of air produced by certain sounds letters make when they are pronounced. Try reciting the alphabet with your hand around an inch from your mouth, and you'll notice puffs of air from certain letters. In English, P and B can be particularly offensive as the air hits the microphone, creating an accentuated boom sound. The shield works by dispersing the air before it hits your microphone's capsule. Where pop shields are not practical, you can use a windshield. These are foam caps that fit over your microphone head. You'll see these on news reporters' microphones when they are outside.

If you're using headphones (which is preferable), don't forget to have plenty of your own voice routed back into your headphones. Hearing yourself is crucial in helping you speak more quietly than you normally would, adding more softness to your voice, which makes for a much nicer, more personal and engaging sound.

I hope that now you're clear on the key elements you need to provide consistency for your listeners and get yourself in a solid routine. Coming up, we will dig a little deeper and look at how you can narrow in and invite the right guests and make them an offer they can't refuse. This is a good moment to remind you why you're doing all this work, and why nailing it is so important: each of your podcast episodes is an asset and acts like a living, breathing business card, advertising your particular expertise and professionalism to anybody who listens. Getting it wrong is not an option!

THE NARROW VIEW

- Consider tone and its effect on emotions as the critical element.
- Create scripts for openers and closers.
- Be prepared for interviews, and remember that you don't always have to be the expert if your guest is.
- Invest in good recording equipment and learn how to set up and position it for the best sound quality.

7

BE MY GUEST (GREAT INTERVIEW SHOWS)

Arguably, this chapter is what this whole book is about. Why? Because great interviews are the linchpin of the entire Profitable-Pod Method, and here's where I'm going to tie it all up for you so you can grow your business with your podcast. Planning a dinner party is all about quality guests. The food helps, and the setting, but ultimately the guests make it a success or failure. And so it is with your podcast. So, what makes someone the perfect guest for your pod? We touched on this in chapter 3, but I'm going to elaborate on not just who they are but also how to get them on your show, and how to take care of them in the run up to and during the show, and, most crucially of all, post-recording.

A narrow podcast leverages relationships with your guests to grow your business. And, at its core, that's what this method is all about. Sure, there are plenty of other things to consider, as this book explains, but without the right guests, you might as well be having a conversation with your lovely but profoundly deaf neighbor. As a reminder of our earlier discussion, each guest you invite should meet one or more of the following criteria:

1. An existing client or potential prospect for you/your business
2. A powerful referrer, either for somebody you want to be referred to, or because they have a network you want access to
3. A potential joint venture partner

When you're first identifying people, start with these three types. Then figure out what is the good, better, best outcome, which we discussed earlier in the book. "Best" could be that they hire you for one of your big-ticket programs that multiple members of their staff get access to. Or maybe you'd rather they opt to retain your services. "Better" might be that they don't hire you personally but have a friend in their (highly qualified) network who would really benefit from working with you. In this case, your guest could make a personal referral, and because they're already in with this person, they know exactly the right time to make that referral. Such perfect timing is usually only something business owners achieve by accident. Finally, "good" is the bottom rung, but it's still positive. It means you've connected with someone who's in your target market or who swims with people in your target market and will talk about you when the opportunity arises. If they have had a great time on your podcast, they will start doing exactly that.

All markets have existing networks of communication. With a narrow position, you don't have to throw your net out wide; you can cast your line and pick out one big fish, or rare fish. Or maybe a handful. Give them a damn good listening to, and then watch as they tell all their peers that you can solve their problems or meet their need.

MEET THE POD-LATIONSHIP CYCLE

Quality marketing doesn't involve scatter guns and hashtags. Rather, it narrows in and creates the right awareness that leads to sales opportunities. It is pretty frictionless if all you need to do is speak twice a month to great people on your podcast, and you'll use the Pod-Lationship Cycle to follow up and keep top of mind.

The Cycle is a system that enables you to manage all the relationships your podcast generates not only in the immediate aftermath of the show but for months, even years, to come; it is the engine of the Profitable-Pod Method. The Cycle can be created and managed in whatever way you choose, but I like to use a Trello board (other apps are available). Trello is software based on the Kanban scheduling system for lean manufacturing, developed by industrial engineer Taiichi Ohno, at Toyota, to improve manufacturing efficiency. At time of writing, a free version of Trello is available that will be sufficient for serving your needs. Trello uses the same Kanban system of cards that track production within a factory. Using the system for your podcast means you can see at a glance where the relationship with your guests stands.

For each potential podcast guest, you create a "card" on which you can note down everything you know about your potential guest, as in the example below.

> **The Cycle is a system that enables you to manage all the relationships your podcast generates not only in the immediate aftermath of the show but for months, even years, to come**

🖥 Leanne Jones

in list Pod Perfect Invitation Pool

≡ **Description**   Edit

Known For / Role:
Former CIO Product development at D-Corp
Now helps scale SaaS subscription-based psychotherapy apps / freelance consultant

Our Connection
We're connected via Alex Adams Email Intro on June 15th

🌐 **Website:**

👥 **Socials**

Insta:
Linkedin:
Twitter:

Other Notes:
Has 7 children.
Loves to sail
Lives in Paris / Atlanta six months of the year
Connected with June Smith who I suspect might be a good fit for my program
May be looking for a community of other SaaS Pros?

W.I.I.F.M:
Best: Hire me and intro to June Smith who is perfect for my group prog
Better: Will speak about me to target market
Good: Shares our podcast episode actively

W.I.I.F.T:
Best: Hires me / Builds relationship capital by referring me to June Smith
Better: Gains profound insights from our conversation - gets a free coaching session
Good: Visibility in a positive light

Fig. 3: Guest card

As you can see, this card contains a short description of the guest that you can add to. The truth is, of course, that not all *potential* podcast guests will become guests, but to give yourself the best chance of getting a "Yes! I'd love to be on your show," the card will really help you. The card contains the key information about the prospect, but it also has two sections at the bottom that are imperative for you to fill in:

1. W.I.I.F.M, or *What's in it for me?* Here is where you note down the good, better, and best results for your business of getting them to appear on your show. Filling this in will stop you from interviewing people who don't have the potential to do the following: a) hire you; b) refer you; or c) partner with you.
2. W.I.I.F.T stands for *What's in it for them?* Turning the first question around enables you to be clear about why they would consider being on your show. The number of podcasts they have been on in the past will determine how easy it is to get a "yes," but what's "best" for them may well be the "best" for you too! For example, if best for you is that they hire you, it should be best for them too, because by hiring you they'll get the outcome they desire. If best for you is that they send you a client, they will also get increased status/relationship capital by referring you to someone in their network.

The people who do the best work in any field, be they plumbers, dentists, accountants, coaches, and consultants, don't need to spend a fortune marketing their business—their existing customers are so happy that they tell everyone about the great work. The main difference for you is that guests might passively share your stuff or talk about you and, if you're lucky, they actively do something about it (either by hiring you or referring people to you). Of course, it's your responsibility to show them how to give you the referral, to hire you, or to partner with you. The follow-up emails I'll share later will present you with various opportunities to lead sales conversations. The other option is they become a partner. Your podcast is a really nice way in because, remember, a beautifully produced podcast is a "credibility enhancing" gift to your guests.

THE RIGHT INVITE

When you invite someone on a podcast in a way that makes them feel important and happy, the reaction is often "Who, me? I'd be delighted. When can you accommodate me?" However, if you ask the very same person to be a guest because you want to "pick their brains" for a few minutes, the response will likely be more along the lines of "I'm pretty busy right now.

And I don't do free consulting." If you're in business, you've experienced this for yourself and know the feeling all too well. I feel your pain.

A respectful invite email looks like this:

> Hi Santo,
>
> I host a podcast called [NAME] about [TOPIC] for [TARGET MARKET]. I'd love to have you on to talk about how you navigated the [SUCCESSFUL RESULT THEY GOT]. You can check out recent episodes in the link below.
>
> What do you think?
>
> Roberto
> Host of [PODCAST NAME / LINK]
>
> P.S.: To lighten your content load, I'll be sure to get the produced episode, show notes, and episode art so you can share with your followers before it goes live.

I appreciate it's not always that easy. Like with dinner party invitations, there's usually at least one person who will want to know who else is invited. It's human nature, right? This form of peer influence is referred to by marketers as "social proof," and it's everywhere you look. Think about the last time you wanted to buy something on Amazon. Two things are now a huge part of the experience. Firstly, the reviews. The same way you're more likely to choose a restaurant with people sitting in it than an empty one, we tend to buy things that other people think are good, even if we don't know those people personally. Secondly, Amazon often recommends products based on social proof: "Other people who bought this item also bought X." Of course, you take a look at that yellow sink plunger because maybe it's something YOU might like too.

Once you understand how powerful and pervasive social proof is, you'd be silly not to use it with your podcast. But the social proof part comes in when you're a little more experienced. The more impressive guests you have, the easier it is to woo new prospects, since you can say, "Look at all the great people I've had on my show before." Of course, when you're starting out, you can't do that. And even for a little while after you start, it might be difficult. But don't worry. There's another way to get to the guests who don't know you. And something that's a huge part of the Profitable-Pod Method. I call it "mentionitis."

MENTIONITIS

"I really like you. Will you be on my podcast?" is a hideous invite. "Be on my podcast, you lucky person" is even worse. Like the template email above, when you make an invite specific, it'll cut through. But here's how to take it a step further. Talk about them with another guest or on a solo show before you make the invitation. For example:

> Dear René,
>
> When you wrote about the power of water in chapter 3 of your latest book, it changed my life. Thank you! It had such an impact on me, and I spoke about that on this short podcast [LINK HERE].
>
> What do you think?
>
> Janet Drinkwater
> Head of H2O Health and Host of Steam Star Podcast

THAT approach is a whole different ball game because if there's something we know about people you're trying to woo, it's that they like talking about themselves. I don't know about you, but if someone emails me and says, "I've been talking about you. Here's a link," even if I don't know who they are, I'm going to be clicking the link at the first opportunity. Letting them know you're already a fan and mentioning something very specific about what they have written or said is a beautiful way to get in with someone. It also shows immediately you're not spamming because it proves beyond doubt you are truly paying attention to them.

Most often, they get back to you to say thanks. Sometimes they ask if you would like to interview them further about their work. If that happens, you're onto a winner straightaway. If not, maybe you need to persevere a little longer. Simply ask them if they'd be open to a conversation on the podcast to share more. And, of course, nine times out of ten, the answer will be yes, because everybody loves talking about what they know. Be specific. Be short. Be polite. Demonstrate authority. And, of course, demonstrate you've already listened to/watched/read their material. Because when it comes to the Profitable-Pod Method, it's all about giving them a good listening to. I use this exact technique personally, as do my clients.

Get a page setup with the right URL linking to the episode in which you have spoken about this pod-prospect. It's incredibly powerful. Positive responses are pretty much guaranteed and, more often than not, they'll invite themselves on before you even need to prompt them!

Mentionitis can also work well in the other direction, and people might be mentioning you. If that happens and you get to hear about it, it makes for a great intro to a pod prospect:

```
Subject: Thanks for the mention

Hi Avneet,

I read that you mentioned me in a blog post
about podcasting. Thanks so much! Would you
like to come on my podcast and tell me more
about how you've applied the narrowcasting
method?

Best, Toby
```

You choose your potential guests based on what's in it for you, but you make sure your invites are based on what's in it for them.

IT'S A YES! SO, WHAT'S NEXT?

Your invite worked, which is great news. And this already puts you way ahead of the game. Now all you need to do is make it easy for them to be on your show.

The first way to make it simple is having a set time of your week or your month when you record podcasts. For example, on Monday afternoons between 1:00 p.m. and 4:00 p.m. By committing yourself to a podcast time, you can set up a diary booking software like Calendly or Acuity to prevent email tennis. This keeps everything looking and feeling slick for your guest.

Of course, for an ultra-high-value guest, you must show up when they're available. There's a balance though. Let them know you can be flexible but, broadly, you tend to record at certain times. Share a link to your calendar, then ask them to let you know if none of those times work for them.

Another way to make your guest happy they said "yes" is being clear with them on what you're asking them to speak about. They will want to look/sound good to their industry peers. But there might be more "touchy feely" stuff that they'd love to talk about. But maybe some of what they do, or are working on, is confidential. The fact is your approach is going to be different depending on who it is and the result you want (good, better, best). In essence, the approach just needs to be a way of getting in with the person and making them look and feel good. Obviously, if it's related to what you do, that's cool, but it's better to let things flow and really just listen to them. Give them the platform to say what they want to. And make sure to massage the ego a little.

With that in mind, ask yourself what you think and what you know they would feel comfortable talking about publicly. By listening to an achievement that they're proud of, for example, you make them feel good about themselves. That's the kind of thing that gets you trust. Then later, post show, you'll have permission to have the deeper conversation.

PODCAST PREP FOR GUESTS

I recommend you have a document ready, something for them to read ahead of the show. Create a private web page for maximum slickness. I normally call the page "book pod." Highly imaginative, eh? But it's not a piece of copy, and it's not a page that's publicly available unless you've got the link (that link being yourwebsite.com/bookpod). It looks like this:

> **Preparation for [SHOW NAME] Podcast**
>
> Not yet booked your interview on [PODCAST NAME]? Schedule our conversation: **Book Now**.

The "book now" button links to your calendar booking system. Once they click your scheduling link, they'll be able to book into your calendar. I use Acuity because not only does it handle all the scheduling, but it also does a few other things like enabling guests to upload headshots and bios so you don't have to chase them for additional content post-interview, and you can get busy creating killer social media assets and show notes.

> **The whole point is that you're going to make them look good.**

The booking form also has a space for them to upload their headshot, which you want for episode art. And there are spaces on the form for them to add a short bio and links to their website and social media handles. The whole point is that you're going to make them look good.

> **About the Podcast**
>
> Thanks for agreeing to be a guest on my podcast for [TARGET MARKET] who want [SINGLE BIGGEST RESULT]. All my guests are brilliant. If you'd like a direct intro to any of those people, please let me know!
>
> The show will normally be live within a week or so of recording.
>
> My goal with the podcast is to build a helpful resource library that all people in [TARGET MARKET] can access.

This part seeds something important we talk about later. And it makes them feel like they're newsworthy, as well as confident it's going to happen. Next, your goal should frame the podcast show to your own business objectives.

> **Where to Find the Podcast**
>
> All published episodes can be found on this page. You can subscribe/follow on [Apple] [Stitcher] [Spotify] [TuneIn] [Google Podcasts].
>
> Schedule your interview slot: **Book Now.**

Everything above in square brackets should be a link. These are buttons. You'll find the official artwork for each podcast platform online. Just ensure when they are clicked, they open in a new browser tab and don't close your page.

Add some visuals to this page with images from the latest episodes. And don't forget to add another call to schedule the appointment for the recording.

You might also want to show them what they can expect from the interview:

> During the interview, I'll ask the following questions:
>
> - How did you get here?
> - What was tough for you?
> - How do you overcome it?
> - What's next?

This style of questioning shows your guest you're interested in them and their story. When it comes to winning sufficient trust to ensure a direct investment in your services or a trusted referral, all of this stuff really counts. The "what's next" question allows them to plug whatever they have coming up for their clients and ideal market. It ensures that your guests are very clear on the fact you are more than happy to promote their stuff.

If you have one, you can also share your signature question on this page. A signature question is something you ask every guest. It's not essential, but it's a pretty cool device that allows for your listeners to anticipate each episode. It also gives you something specific to anchor the episode around with your guest.

Examples:

- Where is one place you think everyone should visit?
- What's the worst journey you've ever had?

One of my clients works in the women's leadership space. Here's what she has on her /bookpod:

Many of us struggle in areas like confidence, asking for what we wanted, stakeholder management, having difficult conversations. We tend to think we are the only ones going through difficulties and we battle on without anyone to talk to. Hearing where you struggled and how you overcame will give other women inspiration and tools to better navigate their journeys.

It might feel vulnerable as a senior leader to reveal that side of yourself ... it's not something we normally talk about openly (especially at work). But remember, I'm asking you to share what you struggled with and how you overcame it—your "scars," not your "open wounds"— and it's the real conversations like this that help other women feel less alone on their journey. Share what you feel comfortable with, and ask me if you want guidance.

I want to celebrate you and raise awareness of your work. Please use this podcast as a marketing asset and tool to connect and reconnect with others. My final question will always be, "What are you working on next?" Please use the podcast as a promotional platform. I'll let my subscribers know what you're up to. And I'll share it on my social channels, I hope you'll do the same.

Let me know in advance if there's anything you particularly want to talk about, or any subject we need to avoid talking about. Podcasts aren't live, so if you say something you'd like to try again, let me know and we'll do that.

Not Booked Yet? [CLICK HERE]

Thanks so much for taking part. If you have any tech issues on the day you can reach me on [PHONE NUMBER].

This example does everything it needs to so that the guest is reassured and has all the information they need to participate. It also has a friendly, generous tone that will ensure your guest feels in safe hands. Tailor your page to suit your style and help the guest. You can even add a little "welcome" video if you want to make it more personal. Done well, it'll show your guest you feel a deep duty of care toward them and create the feeling of excitement to speak with you.

TECH PREP WITH GUESTS

I've already talked about setting yourself up to get good sound, but it's a good idea to take a little time to get good results from your guests too. The COVID-19 pandemic has really helped people up their tech game; nevertheless, it's probably worth sending a short email to help them prepare for the conversation with you. Contained within the body of your email, give them the following instructions politely and respectfully, knowing that they may be a seasoned podcaster:

1. Where possible, please choose a quiet environment for the recording. We would like to avoid editing around sounds created in the home or office setting, including fans, open windows, and passing noise.
2. Avoid environments with too many hard surfaces. A small room with soft furnishings is better than a corridor or large, open-plan area where sound can bounce around.
3. Please make sure you have a strong internet signal where you are.
4. Please ensure all your devices are off or set to silent. If you're recording via your computer, please exit all applications you're not using (email notifications can be especially disruptive). Closing all unnecessary computer programs also ensures your computer power can be dedicated to capturing your voice.
5. Rather than using your computer's microphone, it's better to use the microphone in headphones, which often come with your smartphone. If you don't have any, it'd be great if you can borrow some. Wired headphones cause less interference than Bluetooth.

In your email, inform them that you will test the sound before the recording begins and that any involuntary noise (coughs, sneezes, etc.) will be edited out. Also provide them with the link for joining the recording!

Usually, your sound will be better than theirs, enabling you to appear to be the expert. It'll sound like a radio call-in show where you're the DJ, the master of ceremonies, and that translates to greater authority in the eyes of your guests.

When it comes to the sound check on the day of recording, I have two key tips:

A. Press RECORD! Some of the best podcasts have never been heard because even the best podcast professionals in the world sometimes forget to press record. For heaven's sake, don't forget to press record. And don't wait for the official "we're starting now" bit. In fact, if you can set your recording software to record as soon as the call is started, that will save you a lot of headache. Write RECORD on a sticky note on your screen if you have to. Beyond that, record everything.

B. Use the sound check to put your guest at ease by asking innocuous questions when you're testing the sound levels. Ask them about what they had for breakfast, their plans for the weekend, or whatever. This will gently encourage them to start talking about themselves.

ZOOM RECORDING: MEET THE TWO UNRELATED ZOOMS!

One of my clients is nomadic and likes to record when she's out on the road. You will probably encounter the same challenge several times, but there are a couple of things you can do if you don't want to drag a whole professional recording setup around with you. But here's where it gets confusing. There are two companies called Zoom, and this book talks about both of them.

- Zoom (zoom.com) is video conferencing software, which is what you need if you're recording using a computer.
- Zoom (zoomcorp.com), however, makes handheld recording devices. Crazy, right?

There are various companies who make all-in-one recording devices so you don't need to cart a computer around. The Zoom H4n Pro is probably the most widely used (although every time I look at the website, they have updated their product lineup).

While these devices have microphones built in, you should be investing in devices with XLR inputs so that you can plug your professional microphone into them.

If you're traveling light, you may consider buying a couple of lavalier microphones, which clip onto your lapel. These are often called "lav" mics (you'll see them being used by TV presenters). Lots of companies make them, including Rode, Shure, and Sennheiser. There are different types, so make sure you get one with an XLR.

If you're going super light, you can even get mics that will connect directly to your smartphone. Just don't forget to turn on airplane mode while you're recording.

There are various other options that will give you a recording setup that fits in a medium-sized pencil case.

A FINAL RECORDING NOTE

If you're using Zoom software, when you record, head to the settings and click the "recording" tab; there's a check box that says, "Record a separate audiofile for each participant." This means Zoom will record a separate audio file for you and your guest.

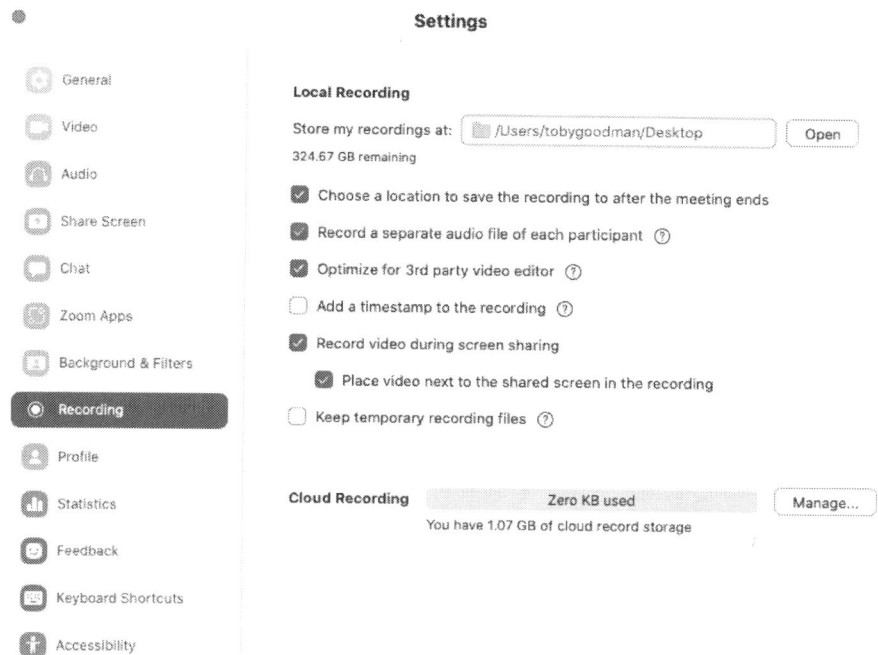

Fig. 4: Recording settings

RECORDING SUPER-HIGH VALUE GUESTS

For guests who can create a massive impact for you, give them the red-carpet treatment. If you've got the option to check in with an assistant, ask them to set up QuickTime and record the audio locally. This is because, with Zoom recordings, you often get a glitch in sound. Even though 99 percent of glitches can be fixed in the edit, QuickTime (free recording software) is a good backup recording tool. You may need to help them set that up, but it's worth ten minutes of your time. Just be mindful that running multiple programs can cause older or less powerful computers more stress and could even crash your recording, so it's an idea to have a fair idea of the age/condition of the computer your guest is using.

Make your guest feel comfortable.

If your valued guest has a poor sound setup, ask yourself what it's worth to have this guest become a client or for this guest to introduce you to one? If you have the dream guest who turns out to be a technophobe, spending $200 on a USB mic, a pair of headphones, and a microphone isolation cube (that will help to block the sound bouncing off walls) is a good investment, and sending/loaning them the kit is a great move. It will make them feel special. Let's face it, 200 bucks isn't a terrible cost for a lead when you look at the lifetime value of that client.

On the actual day, get set up with a good 30 minutes to chill with your high-value guest before recording. You'll want to thoroughly check the levels and that things are sounding good. And then mostly reassure your guest. Fortunately, podcasts are not live, so they can feel at ease to not worry about mistakes, weird noises, or needing to redo anything. The key thing here is to make your guest feel comfortable.

Audio engineers love it if you clap before a re-record because the spike in the audio waveform will show them it's something they need to look out for.

Even though they probably won't be with you in person during the recording, you can speak to your editor during your recording if you or your guest would like to rephrase something. INSIDER TIP: Audio engineers love it if you clap before a re-record because the spike in the audio waveform will show them it's something they need to look out for. For example: CLAP! "Iris! I'm going to ask Rex, our guest, to repeat that last bit."

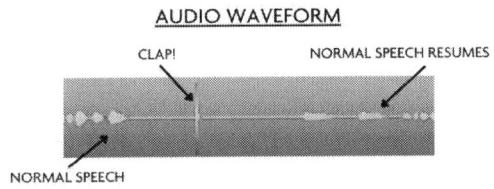

Fig. 5: Normal speech vs. loud clap

WHEN THE TABLES ARE TURNED

Finally, I want to give you a little advice for when you are invited to appear on someone else's podcast (perhaps one of your former guests). Please be selective and don't say yes to everything, especially when there's no evidence of alignment and quality production values.

Also, to drive traffic to your business, create a landing page on your website for podcasts you're going to be on. On this page, you can create a special offer. This could be as simple as your lead magnet, or it could be more unique. Either way, it'll allow you to track the success of being on someone else's podcast and give you more of a chance of the host's listeners identifying themselves to you. Keep the URL simple. For example: yourwebsite.com/name-of-other-persons-podcast

---

THE NARROW VIEW

- Never lose sight of the essential criteria each of your potential guests should meet before you invite them.
- Create a Kanban style workflow to track your guests through the Profitable-Pod Method.
- Document specifics about potential results of being on your podcast for you and your guests.
- Create elegant and thoughtful invites.
- Use the Mentionitis strategy to help grow your network.
- Create a landing page to prep your guests for best results.
- Check your tech is working before you press record.
- Be flexible and keep your communication human.
- Create a landing page on your site for podcasts you guest on.

8

EDITING FOR SUCCESS

Technically, editing is a postproduction process, but here we will consider it part of the production, since a podcast is created in the edit (as are novels, radio shows, TV shows, and films).

One of the great things about hosting your own podcast is the responsibility of great content doesn't sit wholly on your shoulders. However, going in with a plan means you or your production team will have less polishing to do. That's easy when it comes to solo shows, where you are reading a script or speaking from notes. But what about a plan for interviews? That's a harder proposition, since you can never be 100 percent sure where the conversation is going to go, despite having prepared interview questions. Being able to self-edit on the fly is a key skill and will keep your guests engaged and in flow when they speak with you.

INCLUSIVE INTERVIEWING

Getting your podcast off the ground and being comfortable with all that's involved will take some time, so early on in your podcast journey, it's likely that your guests will include people you know already. Indeed, I strongly recommend your first batch of invites go out to the low-hanging fruit. Professional friends and acquaintances will be happy to support you with an hour of their time. (While many of these people may never turn into clients, the chances are they do hang around people who could be. So, these interviews are worth doing.)

As a podcaster, however, you need to be equally comfortable interviewing people you know and don't know. Oftentimes, it's far easier to interview those you don't have a relationship with because your position will be closer to your listeners, and you will likely ask the questions your listeners want the answers to. The danger in knowing your guest too well is that your conversation becomes too "chummy," and that can push your audience away. How do you avoid having a conversation that makes sense to listeners on the outside? This challenge came up when I started working with my client Kathy.

Kathy Sullivan is a remarkable person. She is also the world's most vertically well-traveled person! The first American woman to walk in space, she is also the only person to have traveled to the bottom of the Mariana Trench (the deepest point of the ocean), and that's not all. In her remarkable career so far, as part of National Oceanic and Atmospheric Administration (NOAA), she has advised various U.S. presidents and has been the head of the team building a new home for COSI, the Center of Science and Industry. Like I say, remarkable. Lucky for us, she has a podcast.

Kathy's approach to podcasting was (as you'd imagine) very thorough indeed. Before we started work, I already knew Kathy was a world-class public speaker, used to speaking globally to rooms of dignitaries. I also knew Kathy was well practiced in the art of being interviewed. Where she lacked experience was being the interviewer.

Her podcast, *Kathy Sullivan Explores*, features solo shows that are beautifully crafted gems, while her interviews are deep-dive conversations with remarkable people she is interested in learning from, including many she has met along the way. Some of these guests she knows well, others not so much.

At the start of Kathy's podcast journey, I brought world-class podcaster and journalist Leo Hornak (who has worked for the BBC, *the Times*, and *the Guardian*) in to help. Having listened to an interview between Kathy and a friend of hers (who also happens to be a nationally known scientist who is no stranger to being interviewed), Leo realized that the real gold in the podcast was their unique relationship. The first thing Leo helped us understand is that if you know the person you're interviewing, you will have unique shared experiences only you can share in your podcast. By being aware of this, you can avoid the risk of creating trite sound bites and create a podcast that no one else could.

The second thing Leo helped us with was creating a feeling of inclusivity. With guests you know well, it's important to draw listeners in with questions that include them and feel part of your journey. In Kathy's case, to help set the scene and bring listeners up to speed in an inclusive way, Leo suggested posing the question, "What is your memory of how we met?" Even where the answer isn't included in the final published edit, the information can be used in a solo opening segment, and it will help you give real clarity to listeners and bring them in.

When thinking about your listeners and how you can include them, the following question can be really valuable: "For those who might not be familiar with [subject matter/technical term/location], how would you explain/describe it?" The truth is, you don't have to be an astronaut for technical language to come up in your podcast and alienate your listeners. Take some time to think about technical language risk and plan a few inclusive questions ahead of your recording. Your podcast will be far easier to edit because of this.

BE PREPARED

When you do not have a long pre-existing relationship with a guest, it is important to be respectful to the person they are when you connect with them. You get to know where they are in their career and life by doing your research or simply asking outright.

But when you are speaking with guest experts who you may not have had time or been able to research, due to lack of information or if there is an awkward flow in the conversation, you can use these open questions to keep the focus on them. Trust me, they will love you for it.

- How has your business/skill changed as you have evolved/grown over time?
- How do you know what to do next?
- What sort of success have you seen after [action taken over time]?
- What are the easily avoidable mistakes you see people make in your area of business?
- What is worth spending time on?
- What can/should be delegated?
- What are your favorite resources? Where do you go to get information?
- What sort of things have surprised you?
- What's next for you?
- What's the best way to get started in your industry?

By now you know how to frame your show and how to dreate a great interview. Next, I'll show you how to put it all together.

BUILDING YOUR PODCAST

This book isn't about how to edit audio because we don't have the bandwidth within these pages. However, you already know that, as soon as you possibly can, you should hire a professional production company because you are (probably) not a sound engineer, a professional content editor, or a specialist in writing show notes.

Your editor will cut certain things, which are considered the "non-negotiables," in order to respect the listener. They will cut the following:

- The waffle around lost trains of thought
- The dead air (the silence that can often occur when you or your guest is thinking about how to respond)
- Filler words and verbal clutter (the "ums" and "ers")

- Coughs, splutters, sneezes, and the unpleasant clicking, ticking, and squelching mouth sounds

This takes time, patience, and a certain ear.

Over at our production company (narrowpodcasting.com), we have a number of clients following the Profitable-Pod Method show-build formula. While every show has a slightly different flow, each has a build guide so that our engineers can quickly get started on the postproduction and each episode can be published in a high-quality and timely fashion. The diagram below provides a visual example of exactly how all the elements of your episode could fit together. (Remember, you'll find the show elements, including opener and closer script examples, in chapter 6.)

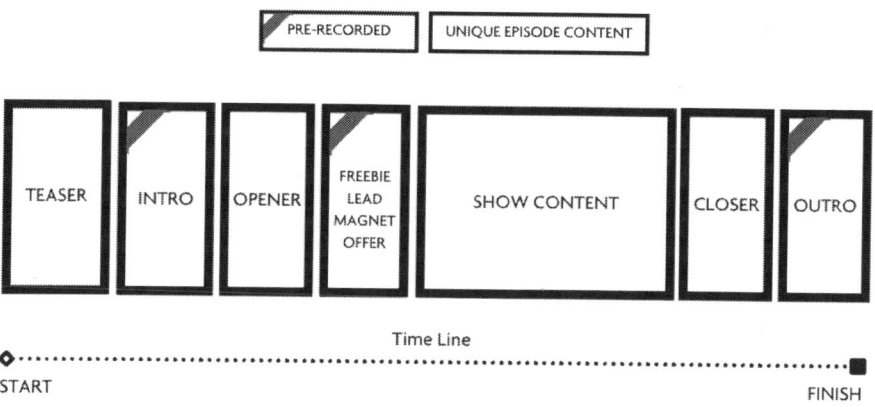

Fig. 6: Building a show from start to finish

20-SECOND TEASER

Your listeners hear teasers first, at the front of each episode, although this teaser will be created at the very end of your postproduction process. This is the unique clip pulled from the episode. If you have a guest on your show, it should always feature their voice. Pull a highlight from the interview that makes people want to lean in. It can be funny. It can be controversial. It can be surprising. It's the first thing people hear and serves to capture their attention. I'd suggest 10-30 seconds is the sweet spot.

If you've ever sat next to someone in a train having a heated phone conversation and wished you knew what was going on at the other end of the phone, that's the feeling we're looking to create here! A client of mine was interviewing a world-famous business coach who mentioned once regretting how much he got upset with his daughter about her outfit one time. This is something that you would never normally hear him talk about, so using that clip as a teaser would make an ideal listener ask themself, "What context did he say that in?"

Try to be aware of these potential teaser moments while you're interviewing someone. If they say something surprising, or unique, or something that you don't think they said anywhere else, look at the recording timer and make a note of the time stamp.

WHAT DO YOU DO WHEN YOU'VE TOO MUCH GOOD CONTENT FOR ONE SHOW?

There will be occasions where you speak with a guest in depth about a number of topics, and even after the audio has been cleaned up, your final edit will exceed one hour. If this is the case, I strongly recommend you create a two-parter. Two individual episodes will give you two digestible shows and more bites at the profitable-pod cherry.

For listener clarity and inclusion, use your openers to signpost your episodes as the first or the second part of a two-part conversation. In most cases, at the start of part two, you should recommend that listeners listen to part one first. At the end of part one, you can also tease what's coming up in the next episode.

WHAT IF YOU MISS A RECORDING/PUBLISHING DEADLINE?

On occasion, life gets in the way and you may fail to record a solo or interview in time to publish in your normal weekly spot. If this happens, it's better to republish an older show. It will continue to refresh the feeds of your listeners who subscribe or follow your show.

Many of your listeners will have discovered your show via a more recent episode, so replaying an early show is a good move. In most cases, you should take a few minutes to record a new opening section so you can explain the situation to listeners and provide context to the old show.

For example:

"I spoke to Robbie this time last year about how he prepared himself to speak online at the last minute due to flight restrictions in his country. This is still a relevant topic for so many of us, so I hope you enjoy this episode from the archives."

This is a quick fix that works great and ensures you retain that top of mind, private slot you get with your listener every week.

But remember, especially in the early stages, this is a fall back and not something I want you to think about until you are a good few months in. And I promise you, once you get into your groove, recording a morning a month, you'll easily take care of all your weekly podcast content for the month ahead. Especially when you have delegated the production to a pro team.

Republishing an episode is also something you can do deliberately (with the bonus of giving yourself a break) if you have an old episode that covers something significant and newsworthy happening in your industry, like a change of legislation. By replaying an episode, you can show your listeners how relevant your content is, even the older stuff, and that they should go back themselves to catch up on missed episodes.

FREQUENCY AND CADENCE

I'm often asked about the rationale behind keeping podcast production weekly. A break might be useful to catch up on recorded episodes. However, if your position is small or micro business owner, or you're new to the space, then publishing weekly shows your reliability and commitment. You're publishing frequently to create a habit in a listener, who then creates space in their diary to listen to your show at a routine time every week. We know that podcasts work best when people are listening at a specific time, and from an "out there" marketing standpoint, nothing is more valuable to your cause. No matter where these listeners are in their relationship with you, nothing is more valuable than owning a piece of their week on a regular basis. That's why I would shy away from seasons, unless you've already got a sponsor in place who is going to fund ten episodes. And if that's the case, the Profitable-Pod Method can still work for you.

Cadence is about intervals. Sticking to a pattern of alternating short, solo episodes with longer guest interviews creates a great cadence. It's easy to record the two short solo shows without worrying about diary alignment with guests, and remember the reason they're short isn't because you don't have much to say; it's because listeners who don't know you yet will only invest amounts of time in you proportional to the amount of trust you've earned. Therefore, ten minutes on a specific subject is more likely to get listened to if they don't know who you are, especially when you've got a very strong podcast name and episode title. The people you interview are going to bring a certain number of listeners to your show. That fact alone creates a transfer of trust and means interviews can be longer.

In short, don't worry about doing this weekly. I promise you, if you plan in advance, you'll be way ahead of the diary, and you'll still be able to take breaks while your podcast does the heavy lifting in your business.

Once your launch is done, you should probably be recording monthly. You might even get it to a point where you record for two solid days, and then you don't have to worry about it for two months!

MORE WAYS TO GET VALUE FROM YOUR EPISODES

In 2004, the standup-turned-podcaster Russell Brand presented a British TV show called *Big Brother's Big Mouth*. This was a spinoff show from the reality TV show *Big Brother* that had Brand speaking/gossiping with a small TV audience about the show. In all, there were 12 spinoff shows created from *Big Brother*. It was cheap TV, repurposing existing content with minimal effort, but it also kept the reality show top of mind.

When it comes to creating more content from your interviews, follow-up commentary episodes are a fantastic way of getting more from the recordings. These will be super valuable if you've had a well-known guest on your show.

My good friend Brady Sadler is the cofounder of Double Elvis Productions, which creates various podcasts about music, including the critically acclaimed *Disgraceland*, created and hosted by Jake Brennan. *Disgraceland's* episodes have a unique take on the lives and demise of well-known music icons, and they now publish *Disgraceland After Party* to the same place they publish the main show. It's a bi-weekly mini-spinoff show that reflects on *Disgraceland* episodes, news surrounding *Disgraceland* subjects, upcoming episode previews, insights into episode influences, listener emails, deleted scenes, and basically anything and everything involving the growing *Disgraceland* universe.

By using the "after party" concept for your solo shows, you can create more awareness and more traffic from your guests without taking any more of their time. For example, let's say you have a podcast about the history of cigars because your business is importing fine cigars for your clients. Let's say you managed to interview Arnold Schwarzenegger about cigar smoking. That's a pretty big scoop for you. Both the people who are interested in cigars and the governor would be very keen to not only listen to your interview but also the short follow-up episode: "5 Things I Learned from Arnold Schwarzenegger about Guilt-Free Cigar Smoking."

This episode (if it were real) would likely attract more listeners to your world and serve as a connection point to attracting the cigar manufacturer you've been trying to get in with since you started your business. Schwarzenegger's ideal reaction would be along the lines of, "You did an episode about me as well as with me? I want your company to import my cigars into Europe."

THE NARROW VIEW

- Be aware of the pros and cons of pre-existing relationships with guests. You'll likely know a few people you interview fairly well—especially when you are getting started.
- Assess "technical language" risk and plan to mitigate it.
- Remember the duty of care you have to your audience and keep your show inclusive by respectfully asking for clarity from your guests.
- Know your show-build inside and out.
- Publish weekly by recording ahead. If you miss a week, put up a replay.

POSTPRODUCTION: PROMO

Big respect for sticking with me so far. Now all your hard work is about to pay off.

Postproduction (in this book, at least) relates to everything that happens after the recording and editing of your episode. Essentially, it involves promoting the show, when the Pod-Lationship Cycle kicks in. You'll be able to do all of this with a few powerful emails a week and track your progress using the Pod-Lationship Cycle. That's how you are going to use your podcast to grow your business.

Post-Coital Pod

One of the most important aspects of the Pod-Lationship Cycle is the "post-coital pod" moment. I consider it the first and foremost part of your postproduction.

If you successfully get your guests to open up and relax, you might be surprised by what they reveal during the podcast. You're showing yourself to be a decent person, somebody who actually listens (and being nice to people is always a good idea), but what you discover could be invaluable when it comes to the post-coital pod moment.

> **At its most basic, post-coital pod is taking full advantage of the beautiful moment after you stop recording.**

At its most basic, post-coital pod is taking full advantage of the beautiful moment after you stop recording. What happens next is something I can pretty much guarantee: they will always ask some version of, "How was it for you?" And that's your first opportunity to say your version of, "You're amazing. Oh my god. It's never felt so good. I can't believe how well you did it. It lasted longer than I thought ..." and so on.

And then because you've done your research and you've been actively listening, you'll be able to open up a behind-closed-doors conversation. This may center around a project the guest mentioned during the recording that might be a possible collaboration. Or you might ask about a connection they have that you would like an introduction to.

If you started your first communication with that level of request, they're going to say no. But after 45 minutes of giving them a good listening to, people tend to be far more open to having those conversations because they understand that you "get" them. Their listening position toward you (the way they listen to you) has changed for the better forever because you are someone who has proved themselves to be a good listener, who is genuinely interested in them. It only works if you really are. If you're not, this won't work.

> **The primary goal is either to have them hire you, refer you, or partner with you.**

This is pretty much what everything so far has been leading up to. In essence, a large part of the Profitable-Pod Method (perhaps all of it) hinges on how you manage this moment and open the door to create more moments after. But none will be more powerful than this moment. It goes back to good, better, best, so choose the right guests in the first place. Remember, the primary goal is either to have them hire you, refer you, or partner with you. But show respect; some people move faster than others. Even if you only get to deepen the relationship in the post-coital pod moment, it should be a win. Because when you choose the right guests deliberately, you can't lose. This is the reward you get for keeping your ears open.

Julian Treasure's books, *Sound Business* and *How to Be Heard*, along with his various TED speeches (viewed by over 100 million people) showed me exactly how I could relate my musical communication skills to spoken ones. He is a real hero of mine. When I invited him on my podcast in 2020, I had lots of follow-up questions because I wanted to know how he thought the pandemic was going to change and challenge how we communicate. Our conversation was natural, and I really had no expectation, but my ears were open for opportunities to serve and collaborate. He talked about podcasting, but because he is one of the world's foremost experts on listening and speaking, I knew he wouldn't need help with that. But somebody to partner with to help produce podcasts? That was something that really excited him. Did I make a deal a few minutes after the interview? Absolutely not. But I was able to gain his trust and book a chat in the following week.

Had I just approached him cold and offered my services, I would've almost certainly been rejected out of hand. But after 45 minutes of conscious listening, he was much more receptive. Because he knew that I understood the value of how he and his business serve their target market, a segment of whom happens to be the same as mine. That is the magic behind a narrow podcast with a good post-coital pod chat. Everything I've shown you so far leads up to this moment. Don't squander it!

Next, we are going to explore the more traditional aspects of podcast promotion.

9

GETTING SHOWS OUT THERE

Back in the winter of 2017, I was stuck in a pit. I was alone and separated off from everybody else. And sometimes it was really dark. This wasn't a pit of despair or any other kind of metaphorical pit. This was an orchestra pit, the kind you get six feet below the stage at a theater. My drums were set up facing outwards, so I was very literally center stage as 2000 people a night came to hear what I helped to create. Nevertheless, I most certainly wasn't the star of the show. You don't have to be either.

A few months before, I had been listening to James Schramko's *SuperFastBusiness* podcast. I had carved out time specifically to listen to him and what he had to say, and I'd learned to trust what he said. After all, he is the man who so succinctly showed me the importance of having an offer that converts.

One week, he had a guest named Matthew Kimberley on his podcast. Matthew was talking about how to sell better and mentioned a book called *Book Yourself Solid*, by Michael Port. I was enjoying listening to the show, when James ended it abruptly saying the rest of the interview (recorded at his in-person mastermind event) was available if you joined his membership program. Now I don't know about you, but I'm often skeptical of these tactics and ALWAYS skeptical when people make offers. But with James and Matthew, there was what's known as a "transfer of trust." I already trusted James, because I'd made time to listen to his podcast regularly and it had really helped me. By extension, therefore, I trusted Matthew, too, and what he said caught my attention as well. Here's the important part though: I didn't spend money with Matthew (the guest). I spent money with James (the podcaster). I paid four figures to get into his membership program and hear the rest of Matthew's talk. I downloaded it and watched it in my pit between shows. What that example shows is that as a host, you'll still get the credit from your listeners even when you're not the expert.

When people talk to me about "getting out there," I immediately glaze over and remember the many badly or unpaid gigs I did as an exploited young musician in the name of a "good opportunity." Then I think of Michael Port's *Book Yourself Solid*.

Most people think the answer is in getting out there. It's not.

In fact, if you know anything about Michael's book or system, a lot of what I've talked about might seem familiar. Reading his book and applying the system completely changed how I thought about everything, and, on reflection, helped me see how I got hired to support some of the best-known musicians in the world. It allowed an introverted drummer like me to grow a profitable podcast production company and consultancy without needing to be center stage and help hundreds of other successful business owners do the same. The same can happen for you in your chosen field with the Profitable-Pod Method.

PERFECT PODCAST PROMO

The vast majority of people spend way too much time and energy "getting out there." They do it with podcasts. They do it with social media. In fact, they pretty much do it with everything in their lives. Our culture is fascinated with celebrities. People look at them and see a carefully curated image of what the celebrities' lives are like. It looks great, and they want it too, so they figure that getting out there will give them some of that fame. And once they're famous, their business will simply build itself. This is not only wrong; it's dangerous. Because you'll most likely end up disheartened and broke.

A lot of people who are not reading this book will think that getting that episode out there is all important. Well, it's not. In fact, it almost doesn't matter if nobody ever listens to your podcast! But for maximum effectiveness and impact for your guests, there is some promo to do.

When it comes to pre-launch/launch promo, a lot of people fall into a trap: announcing a guest ahead of the recording. But, if you're targeting the right guests for the right reasons, this is usually a mistake. The episode is all about the GUEST. You're trying to woo *them*, not the listener.

Remember good, better, best? You have a desired outcome in mind, and your job at this stage is to make the guest feel good. To do that, you take the custom episode artwork you've done, the show notes, and a little tagline, and you create shareable assets.

> **But you don't announce the guest until the episode is "in the can."**

But you don't announce the guest until the episode is "in the can."

Why keep it close to your chest? Sometimes people miss their date with you. Sometimes they reschedule right away; oftentimes they don't. If you've done your homework, the release of your episode should coincide with their new thing, be it a book publication or a product launch or whatever they're promoting so that they get to feel good about being your guest. So, you take the assets you've created, and once the episode is in the can, you use the Pod-Lationship Cycle to share them in a specific way, at specific times.

This basic level of episode promotion does a couple of things. It demonstrates your professionalism, and it can help build authority in your market, which is very different to the kind of broad celebrity status many people chase. The absolute must-have are show notes. While their primary purpose is to help get your show indexed, if done well they can be very useful to your promo.

I recommend using content from your podcast and turning it into three social media posts (at a minimum):

1. The first one features episode art, a headshot of your guest, and the title of the episode. Easy.

2. The second one requires the show notes, as it features a quote from the show. Something easily memorable your guest said.
3. And the third is an audiogram with a 20-second (or thereabouts) clip from your podcast. This can be the episode teaser or a highlight from your show. A professional production company will create show notes you can take quotes from and pull-out audio clips for you.

These three things are the minimum you need to do when it comes to episode promo. I'll show you when and how to use these three assets in the next section, where I'll finally reveal the magic and simplicity of the Pod-Lationship Cycle. In the meantime, I'll tell you about two tools my clients use to get this stuff created.

Canva.com

Canva is a simple online graphic design platform designed for people who aren't graphic designers. You can use it to quickly create various sizes of artwork, from business cards to logos. Once you have set up your podcast album and episode artwork, you'll quickly be able to duplicate episode art and "quote art" (short, eye-catching quotes from the podcast) to publish on your website and social media feeds. More importantly, you'll be able to share them with your guests.

Wavve.com

When it comes to creating audio posts, you need to create a graphic that leaves space for subtitles. Then you can move it over to a tool to create audiograms via Wavve. These are short videos that display subtitles and play audio.

Creating this stuff is easy enough, but it does take time. Over at the production company, we created tutorials for our clients who follow the Profitable-Pod Method. Most have been able to share the tutorials with a team member or virtual assistant and delegate the creation of the assets.

SEARCH ENGINE OPTIMIZATION: DON'T FORGET YOUTUBE!

The next step is SEO.

Podcasts are available everywhere. You can find them on Apple, Spotify, Stitcher, and a whole bunch of other places. And the funny thing is, some podcasters think it's a good idea to tell people that. When you overwhelm people with choice, they don't take action.

But when is a podcast episode not a podcast episode? When it's a YouTube video. YouTube is one of the most underrated channels for podcasting unless you're Joe Rogan, which you're not. But you don't have to video all your podcast interviews to make use of YouTube. Strip the audio out and put a picture up (episode art), that's enough. You can use the automatic transcription/subtitling function inside YouTube so that the content is searchable. There's a standard format for subtitles called an SRT file, and of course, a file with words is indexable (by any search engine).

YouTube is the world's second largest search engine after Google's main search engine, so you're getting even more traffic to your podcast. You can also use the same method for Facebook, LinkedIn, and so on. Only you know where your customers hang out. I encourage you to test and measure these approaches to make sure you know where your customers are.

The audio is enough, but the show notes that get loaded into the description help your podcast get found on YouTube. If they don't want video, they will be able to link to you directly and subscribe according to their own preference. Use the power of YouTube/Facebook/LinkedIn and so on to carry people over to your podcast. Where can your podcast be found? Everywhere! But all the links are over on your website. This is good for you, especially when outages on social media platforms are not unheard of.

ACCESSIBILITY: MORE ON INCLUSIVE PODCASTING

There is another strong reason to be on YouTube, Vimeo, and other free video platforms. By using video, any podcast followers who need to read along, rather than listen, can access the content. This doesn't mean you need to start filming. You can use a simple piece of episode art, resized in a 16:9 ratio, with subtitles at the bottom.

When it comes to YouTube and even Facebook, you can opt for having subtitles automatically generated (they aren't always spot on), or you can use a transcription service such as rev.com or trint.com to get more accurate human transcription (this shows you're committed to full accessibility). It also widens your potential audience.

Incidentally, while we're on the topic of accessibility, it's important that the website you design can be easily read across all major browsers and renders well in mobile format.

OWNERSHIP: AN IMPORTANT REMINDER

It is essential that you own your podcast because it's the source that feeds the other places your content appears online. To do that, it's essential you invest in hosting, which I encouraged you to do way back in chapter 1!

Podcast hosting platforms are the tech that turn your audio files into actual podcasts by creating a unique RSS feed and sending the content to all the places podcasts can be found. Without proper hosting, you don't have a podcast.

Todd Cochrane is the founder of Blubrry Podcasting, which hosts over two million of the world's independent podcasts for its clients and has been instrumental in helping my production team ensure the podcasts we're producing for our clients are safe. Please, please, please make sure you invest a few bucks a month in podcast hosting and avoid anything "free." Although it seems tempting, free podcast hosting companies can essentially hijack your content and fill it with adverts and all sorts of other content to promote their own agenda. If you're looking for proper podcast hosting, head over to narrowpodcasting.com and I'll hook you up.

Once your podcast hosting is set up, you'll be provided with a small bit of code, often called an embed code that you can paste into your specific episode blog. Once applied, the podcast player will appear on your blog page. This means people can listen on your website too.

CREATE SHOW NOTES

Episode show notes (often referred to as just show notes) are notes about your show that you should upload to two places: your hosting platform and the blog page you create for each episode. You need to put them just below the aforementioned podcast player on each episode blog page. Show notes should be easy to read and focus on SEO so that your show gets indexed high up in search engine rankings. When people are searching online for content related to your show, your show notes text gives you a good chance of getting found.

Show notes are not and do not need to be blog posts, but they can serve as the basis of blogs. I often refer to them as "fire insurance level repurposable content"! That's because they'll really help you when you start creating social media content and emails.

Just like audio editing, well written show notes require time and skill; that's why we always include show notes in each podcast production package.

The content for show notes includes the following:

- Podcast title, episode number, episode title
- Short bio of the guest
- Episode description: overview of what was talked about
- Episode bullets: summary of key talking points (encourages the listener to listen)
- Quotable quotes: words of wisdom from the guest (sprinkle them throughout the notes or pull them together)
- How to find out more about the guest
- Links to products/services/resources mentioned by the guest
- Call to action: how the listener can connect with you/your product or service
- Call to action: encourage sharing, liking, and subscribing

Here's an example:

> DESIGNED FOR LIFE PODCAST | Episode 05 - Renew Your Career (and Life) with Life Design Expert Mike Smith
>
> Mike Smith is a Career Coach and Career Design Specialist at [Business Name]. [Business Name] offers advice and coaching that specifically caters to those who are in their second half of life. He has worked as a programmer, engineer, marketer, and IT consultant for large companies like IBM and Apple as well as startups in the tech industry. Mike is also the author of the book [NAMED LINK] and hosts a podcast with the same name.
>
> Mike joins me today to share his passion for guiding others in their second half of life. He explains the necessity of planning your life beyond retirement age and why having a career is not out of the question. He discusses the concept of his book and what to expect from the third edition. Mike also describes how networking can help you boost your career, especially in midlife.
>
> "Our world has changed. We must prepare for what's going to happen next." - Mike Smith
>
> Today on [SHOW NAME]:
>
> - How Mike helps the generation of baby boomers
> - Why he wrote his book [BOOK NAME LINK]
> - The difference between money spent on healthcare in the U.S. and Mexico
> - How longer lifespans changed the way we look at the future

- The similarities between younger and older generations in terms of career demands
- Helping those who need advice with breakdown
- How networking can be used to your advantage in this stage of life
- What ABC stands for and how it improves your career
- Why the best decisions he's made are the ones that caused him to fail

Mike Smith's Words of Wisdom / Tweet-ables

"Push yourself, because no one else is going to do it for you."

"Sometimes later becomes never."

"Great things never come from comfort zones."

Connect with Mike Smith [Website Link], [New Book Launch Page Link], [Book 1 Link], [Link to Mike's Podcast]

Get FREE Access to [SERVICE or PRODUCT]

Are you struggling with knowing what, when, and how to focus your efforts to get things done? As a long-time sufferer of severe procrastination, I understand the struggle, and that's why I created the [Course Name] just for you. This system has helped thousands like us clear the clutter and focus energy on getting the right things done.

The [FREE COURSE LINK] is based on my proven system to help you focus your attention and maintain momentum around your day-to-day life. Sign up for the [FREE COURSE EMAIL SIGN UP LINK] today!

> Thanks for tuning in to today's episode of the PODCAST NAME: SLOGAN with HOST NAME. If you enjoyed this episode, subscribe to the show on [PREFERRED PLATFORM] and leave us a review. The show can be found anywhere else you may wish to get your podcasts too!
>
> Be sure to visit [WEBSITE] and connect with us on Facebook, LinkedIn, Twitter, and YouTube, and if you know someone who you think would enjoy this episode, please share it with them on social media. [LINK TO SHARE]

PODCAST NETWORKS

The Profitable-Pod Method doesn't require you to join or start a podcast network. But it's good to be aware of them and how they work, because one day you might be in a position to join or start one. I'm arming you with the following information so you don't fall into what I like to call "the network trap."

There are two types of podcast networks. A podcast network is (should be) a group of individual podcast shows whose shows share common interests. They make money by cross promoting and sharing the same advertising space. Many even create in-person events to serve the audience. Built and run correctly, podcast networks generate credibility, community, and referrals, as well as build audiences and customers fast!

Then there are the bad podcast networks. These nefarious networks are set up with one goal only: to use other people's content to make money for themselves. You can spot them a mile off because they don't have a clear theme to the shows. They use the rule of primacy and recency by bookending each show with the network name to get all visitors to their own website, not the shows' creators. They exist because they use the shady tactic of inviting people to join the network.

THE NARROW VIEW

- Create show notes with SEO in mind by thinking about what ideal clients would search for.
- Create 3 social media posts from your title, show notes, and audio for each episode.
- Don't forget to upload your podcast to YouTube; it's a very easy win.
- Use a professional podcast hosting company so the ownership of your podcast is protected and your source content is safe.

10

AMP UP YOUR NETWORK (GROWING YOUR AUTHORITY)

Authority is hardwired into the human brain. In July 1961, Stanley Milgram designed an experiment to test this. He asked volunteers to administer electric shocks to strangers when they answered questions incorrectly. Every wrong answer increased the voltage applied. An authority figure was present to test how far the volunteers would go and encourage them to continue when they became uncomfortable.

Milgram changed the way we think about authority forever. In the experiment, the volunteers went against their own internal moral compass simply because there was somebody else in the room that they perceived as an authority. The experiment has remained controversial for the past 60 years and had a huge impact on criminology and psychology. But since then, more and more people have started to look at the implications for marketing.

Authority can be conferred by social norms, for example, by the police or courts of law. But authority can also be built by anyone. We all have "that friend," the one we go to when our computer is broken or need help shopping for clothes or fix something in our house. These people have built authority with us over a period of time, and they're part of our network. But we often don't think about it in those terms, but YOU need to get a little more deliberate about this.

The first (and perhaps not obvious) step to growing your authority is to be excellent at what you do! Then people need to know that you're excellent. And then they need to remember you exist.

Because we're all busy people, more so now than ever before, I like to keep in touch with good people I know using a method adapted from Michael Port's advice in *Book Yourself Solid*. There are few ways to implement it, but the basic idea is to put a system in place to ensure that you're consistent in your networking. However, that doesn't mean consistent in your *hustle*, because that's a turn off. For example, here's how NOT to network:

> Hi Jane,
>
> You good? I'm working on my NEW PROJECT, and you can BUY MY BOOK. You can listen to MY PODCAST, or you can BUY MY COURSE.
>
> Thanks,
> Bob (pushy) Smith

Not the best idea. I get emails like this all the time, and I imagine you do too. They make me feel uncomfortable, disrespected, and, on occasion, violated. And seeing as we are all in the business of creating feelings, this is never going to be a good way to go. Very rarely do we get emails we enjoy reading, and the truth is, you don't need to be an incredible writer to connect with someone over email. You just need to show interest and respect people's time. A little knowledge doesn't hurt either.

My Pod-Lationship Cycle notified me that I hadn't been in touch with Anna for a while. Thinking about Anna, I had no set agenda other than to simply keep in touch with someone I respect and haven't spoken with in a while. So, I sent this:

> Hi Anna,
>
> Turns out it's been three months since we spoke. How are you doing?
>
> Toby

I try to send an email like this most days (obviously to different people) because it's good to keep in touch with good people. She's not in my target market, but being someone with a significant business, some of her clients *are* in my target market. We are not competitors. As a highly respected therapist, she's not about to dish out podcast strategy advice to her patients, and I am certainly in no position to do what she does. Her reply came a day later, and it took me by surprise.

> Toby!
>
> Great to hear from you. Family all good.
>
> I've just been speaking with a client. I think you can help him. Do you have space? Can I make an intro?
>
> A x

You can guess the rest. If you're wondering how I met Anna? I interviewed her on a podcast.

It's nice to be nice, and it's okay to have a system in place that means you are top of mind every few months or so with other professionals. It will also really help you grow your business. With ninety people on your private email list who you know are good professionals, you'll be able to touch base with them once every three months or so. These are people you might be able to (and should be prepared to) help at some point, or perhaps might be in a position to help you. To make this work, you must curate the list.

Do not put these people in an email program. Send short, personal notes from your email address. Trust me; it is worth the time.

Once you get in the habit of connecting in a genuine way, it increases your presence. And presence gives you more opportunities to demonstrate your authority in a very cool and passive way. If you're an authority in your market and you're top of mind, your business will grow. Thank you, Michael Port!

But how do we apply this to your podcast? I'm glad you asked ...

STAY TOP OF MIND: THE FOLLOW-UP

Once somebody has been a guest on your podcast, they've seen you demonstrate authority. And they've also experienced you giving them "a good listening to." We've already talked about post-coital pod, and what a crucial part of the method it is. But how do you stay top of mind? Well, the other thing that's essential (like with any good marketing system) is follow-up. Because, at the end of the day, sometimes it takes a while for people to realize what a great experience they had with you. Not least because they may need to get off the recording with you to attend to a pressing work or personal matter. The following follow-up system works a treat for guests who have recently been on your podcast.

Why Follow-Up Is Essential (for Guest and You)

This is about what happens at the end of the podcast. Simply, you're going to send four emails.

Email 1 - Thanks for your time (sent immediately after the interview).

Email 2 - Let them know when the podcast is about to go live and include episode art.

Email 3 - Let them know how it's been received and include quote post.

Email 4 - Let them know you still think of them often and further prove your value with intros, include audiogram.

The person you invited onto your show is not the person you are now emailing in the sense that you now know them better and there is a bond of trust. They know you're smart and you care about them. You've let them talk about themselves. And they're buzzing from having talked about their success. So that moment is a golden time for you. And it's because of this, your guest is likely to be far more open for more intimate and private conversations about one or more of the following: a) How you might be able to help them, b) Who they can personally introduce you to, c) How they might be able to partner with you.

So, that's why it's important to remind yourself of the good, better, best outcomes for you and your guest. When you know that, you know what an effective post-coital pod conversation looks like. And after that, how can you continue that conversation? Send them four personal emails after your recording.

This first email should be sent directly after your chat. It prolongs the warm glow. If your post-coital pod moment went well, good things may happen fast. Either way, it's professional and polite, and that's the main objective here.

Email 1: Thanks so much [NAME]

Dear [NAME]

Thanks so much for your time just now. I'll send you links and a few social media assets as soon as my team has them made up.

Warmest,
[NAME]

PS: Would you like to schedule a time to connect about [SOMETHING MENTIONED IN THE POST_COITAL POD MOMENT]? How's 2 p.m. EST next Wednesday?

IMPORTANT! This email is all about being respectful. It should also capitalize on (in an appropriate way) what happened in your post-coital pod moment. The first part is short, polite, and respectful. The second part where you promise to share links will get them excited about hearing the conversation back. The "PS" is where you get to make an offer. This offer might relate to a challenge the guest has shared, for example, an employee who may need some coaching.

Sense check! You don't need to jump in and make an overt offer. In fact, sometimes, you may need a few moments to think about your PS. This book is about harvesting and creating long-term, strong relationships, so right now is a good time to remind you (and me) that, contrary to how social media can make us feel sometimes, not all problems can be solved by a pithy one liner or meme. Equally remember, the PS could be asking for an intro to someone you know your guest knows, or it could be a further discussion about a partnership opportunity. If you truly are stuck and simply not sure, you can ask them if there is anyone else you think would be a good guest.

Important reminder. You must publish your podcast to a unique blog page on your website. If you miss this step, you lose out big time. The URL should be yourwebsite.com/podcast/episode-name-with-guest-name. When that is done, it's time for email 2.

Email 2: [NAME]! Your podcast interview goes live tomorrow!

Dear [NAME]

Thanks so much for your time [WHENEVER]. Your show goes live on [DAY/DATE/TIME/TIMEZONE]. Here's the link: [yourwebsite.com/podcast/name-of-episode-with-name-or-guest]

Please note: this page will only go live at the above time.

Can't wait to hear what you think.

Warmest,
[NAME]

PS: To help lighten your content load and keep your followers up to date, I've attached episode art for you to share. Please tag me on the socials when you post. Thanks!

Send the following email ONE WEEK AFTER SHOW GOES LIVE.

Email 3: Feedback on [EPISODE TITLE].

Hi [NAME]

Thanks again for your time on the podcast. Your episode has been live for a week now, and feedback has been really positive.

I loved what you said about [INSERT RELEVANT TOPIC]. In fact, I thought it was so poignant I had a shareable quote image made (see attached).

How are things going with your [LATEST PROJECT]?

Warmest,
[NAME]

PS: The full link to your episode and show notes is here: yourwebsite.com/podcast/guestname

If you don't have knowledge of their latest project, you can ask them about what they're developing now. This information might later be useful for a return podcast invitation.

Send the final email in the series three or four weeks after episode goes live.

Email 4: [NAME]! A gift.

Hi [NAME],

Can you believe a month has flown by since your show went live? The team has created an audio post for your episode, and I thought you should have it too. There's nothing like the sound of the human voice to cut through the noise out there.

Warmest,
[NAME]

PS: Is there anyone I've spoken to on the podcast before or since your episode you'd like an intro to? If so, I'd be more than happy to connect you to the podcast alumni! You can find the full list over at yourwebsite.com/podcast

All four emails give you a reason to connect with your guest. They all build momentum. While the first three emails are focused on giving social media assets as gifts and keeping you top of mind, the final one is a gift of your richest asset and introduces a new way to build your connection and relationship.

The final email positions you as THE person to know in their network. You're giving the most impressive and remarkable social media asset to them in the form of an audio post. Secondly, you are building your relational capital by offering to connect them personally with someone you know and suspect they want to know. To do that, they have to go back to your website too, and that is never a bad thing.

ALUMNI CONNECTION

This is the final part of the Pod-Lationship Cycle. All guests that have appeared on your show are now alumni. So, start connecting them! This alumni play is a big deal for you and really helps establish your authority.

Before I go on, a warning:

For clarity, you should NEVER ask for or expect any sort of referral fee. That's not even slightly cool. Good karma is where it's at, my friend. The last thing the Profitable-Pod Method is about is bait and switch, and while there are lots of podcasters who reel in guests with an invite only to send an invoice for "free marketing," the Profitable-Pod Method is refreshingly different. Good business means a win-win for all parties.

Look out for the people in your network. Short introductions are a great way of creating abundance for people who have given you time—and they create good karma! Be intentional about not letting others in your network leave money on the table by sending an email like this:

```
Hi [NAME],

I don't know if you know [PODCAST ALUMNUS],
but it occurs to me you're both working in the
[PRODUCT/PROJECT/MARKET] space. You both
spoke with me about significant challenges
you've had on the podcast. So, if you aren't
already connected, I have a strong hunch you
should be. Let me know if you're open to an
intro, and I'll make it happen.

Warmest,
[NAME]

PS: You'll find [PODCAST ALUMNUS] episode
here: [yourwebsite.com/podcast/podcast-
episode-with-name]
```

Now you're being the connector. You want to connect those two people because you're a thoughtful person and want to be talked about more in a positive light. By connecting intentionally, you can attach a link to something that proves the connection, ideally a podcast episode. This immediately and dramatically increases the chance of getting them back into your world listening to you and asking questions. Alumni networking positions you as a fiduciary who can be trusted and relied upon.

Be prepared for how things could play out when making introductions. Here's what happened when my client/podcaster, Sarah, sent her guest, Rob, a professional event planner, an audio post. Rob saw that she'd interviewed Carly, a baker, in a recent podcast episode.

> Sarah!
>
> Thanks so much. Love the audio post. OMG! I didn't know you knew Carly from Carly's Crazy Cakes! Would love an intro.
>
> Thanks,
> Rob

Sarah had interviewed Carly but didn't know her well, so she needed to check she was open to an intro. So, here's what Sarah sent back to Rob.

> Rob!
>
> My pleasure. Glad you liked the audio post, please do share it with your network. I'd be happy to ask Carly if she's open to a chat with you. I know she is super busy, like you. Is there anything specific I can say when I ask?
>
> Best,
> Sarah

If Rob didn't reply to Sarah, she used the Pod-Lationship Cycle to remind herself to follow up a week later. Here's what Rob sent back to Sarah:

> Hi Sarah,
>
> Yes sure. I'm a huge fan of Carly's products and would like to find out if she'd be free to bake for our best client's Christmas party.
>
> Thanks,
> Rob

Sarah then emailed Carly.

> Subject: Intro to Rob Green?
>
> Hi Carly,
>
> I've just been in touch with Rob Green who has an event coming up that I think you might be a good fit for. He was also on the podcast you can check out the episode here: yourwebsite.com/podcast/robgreen. Let me know and I'll make the intro.
>
> Warmest,
> Sarah
>
> PS: Let me know if there's anyone you'd like me to connect you with. Here's a link to the people I've spoken with on the podcast. [yourwebsite.com/podcast]

When Carly responds, all Sarah needs to write is this.

> Subject: Carly meet Rob
>
> Hi both,
>
> Making the intro as promised.
>
> Warmest,
> Sarah

That's it!

MATCHMAKING

Back in the 1980s, I used to watch a TV show called *Blind Date*. Host Cilla Black introduced three singles to the audience. They were then asked questions by the potential date, a single individual, who could hear but not see the other three, to choose with whom to go on a date. The couple picked an envelope and discovered where they were going on a date. The following episode showed a short clip of the couple on their date, and Cilla would then sit on the couch to speak with them to find out how they got on. This is the kind of matchmaking I encourage you to do!

When you have made the match but don't hear anything, set yourself a reminder to follow up one or two weeks later. Even if nothing works out, it's a great way to prove you're a decent human and to remain top of mind. While these are far from blind intros, send individual emails, just in case things didn't go smoothly. In Sarah's case, here is what she sent to the individuals separately.

> Subject: How did it go with Rob/Carly?
>
> Hi Rob/Carly,
>
> Just checking in. Did you manage to connect? How did it go?
>
> Best,
> Sarah

Ideally, you find out and things worked out well. Even if they didn't connect successfully, you were the one who made the effort. This sort of behavior increases the likelihood of these guests asking you questions. If they are thoughtful human beings, they will be wondering what you're up to now and how they can help you. Of course, this is all about where they are in their relationship with you. Once you've had a great chat with them, put the podcast out into the world, done the show notes, sent them an image of their face around your logo, and so on, you get to ask those questions, and they're likely to be answered. When it comes to your business, that's one of the most powerful positioning tools you can ever have.

USE YOUR LEVERAGE

Two more simple ways to leverage your podcast to grow your network and authority.

1. Retain clients with a podcast

You can do this with solo episodes as well as guest episodes. You send a specific episode to somebody, and you can tell them that you created the episode with them in mind. ("This podcast episode is perfect for you. Haven't heard from you in a while. Are you still struggling with X? Are you still working on X?") Sending a specific episode to people for whom it might be relevant is powerful. But it gets even better if you know current clients are struggling with a particular issue. When you have an episode that you know touches on that problem, sending it is almost always well received.

Beyond that, of course, it positions you as the authority. And that leads to retaining clients and staying top of mind in between client meetings.

2. Create clients from listeners

Of course, the same thing applies to getting new clients. You'll have people in your network, email subscribers, or just people on social media who ask you about certain issues.

Sometimes, it comes in the form of them wanting to "pick your brain"; sometimes it's a little more subtle. Either way, being able to give them a laser-targeted prescription is not only less work for you, it's much better positioning. If you're good at what you do, you absolutely should **not** be allowing people to pick your brain. Hopefully you know that. The beauty of a profitable pod is using it as a resource to not only build authority, but genuinely help people.

You can do exactly the same thing with a specific episode. And for some prospects, this will bring them close enough to actually want to work with you.

POD-LATIONSHIP CYCLE IN ACTION

Ultimately, all of these strategies are designed to make the absolute most of your podcast. You're building your relational capital by growing your network, growing your authority, and being the go-to person for everybody you connect with—one of the best positions you can occupy as a business owner.

In chapter 7, I introduced you to the Pod-Lationship Cycle, a system that keeps me and my clients on top of choosing ideal guests, inviting them, recording the actual show, and then tracking all of this follow-up that leads to clients. As I mentioned, I use a free Trello board (other online platforms are available) to give me a dashboard view of my podcast and where I am with everyone, from possible guests to all people who have been on my show. Moving from left to right, here's what it looks like:

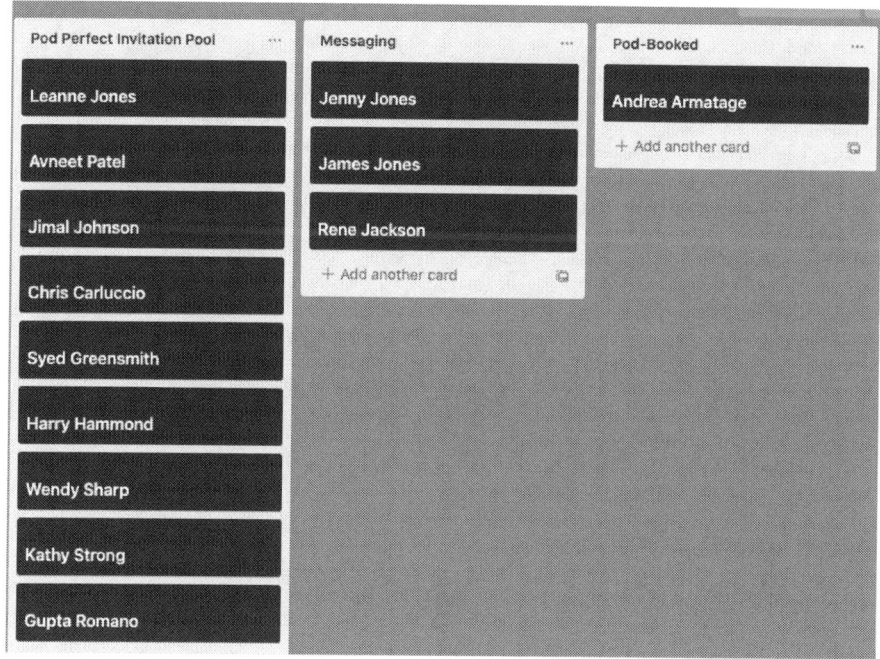

Fig. 7: Pod-Lationship overview from pool to booked

Pod Perfect Invitation Pool

Put all your dream guests here. Don't forget to create a card and populate it with information about the guest as I mentioned back in chapter 7.

Messaging

Move the card along here as soon as you get a response from the prospective podcast guest. You're in communication, but nothing has been agreed yet.

Pod-Booked

As soon as you've booked your podcast recording, move the card here.

Following the podcast recording, the Pod-Lationship Cycle really kicks in with the four emails I described earlier in the chapter. The Trello board now looks like this:

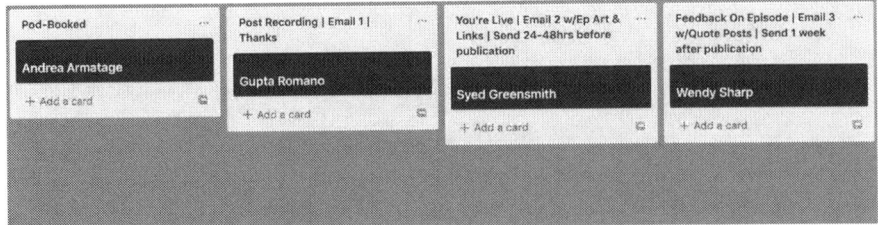

Fig. 8: Pod-Lationship overview from booked to feedback

After the four emails have gone out, then your contact will be quarterly, or as matchmaking opportunities arise. Now you're into the personal keep in touch strategy. You can add comments and set reminders on your Trello board (or whatever you use) so you can keep track of where you are with your relationship as it progresses. Use your Trello card to make notes and set reminders, as below.

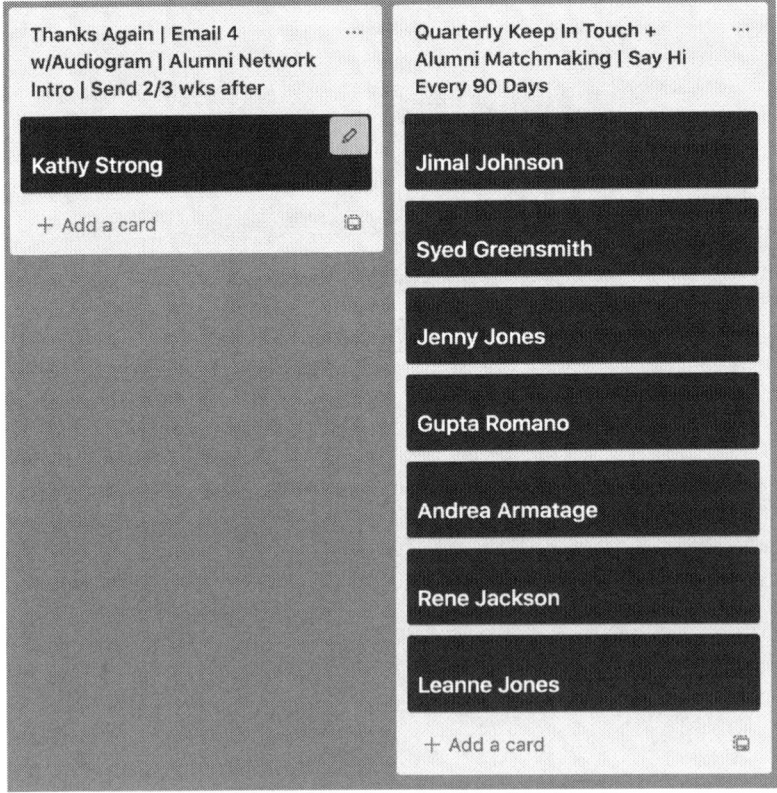

Fig. 9: Pod-Lationship overview, including "thanks again" and quarterly follow-up

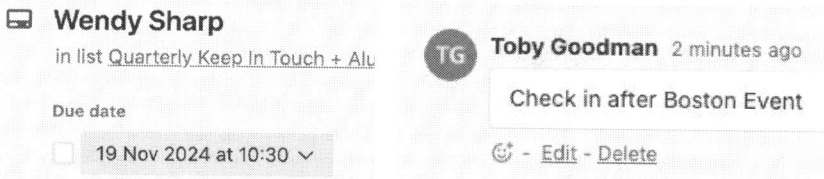

Fig. 10: You can make further notes to remind yourself why you set those reminders too!

TRACKING SOLO SHOWS

Using this same system to plan and track your solo shows and how you use them is pretty simple. Just set up two columns. Fill the first one with show ideas and the second with shows you have produced.

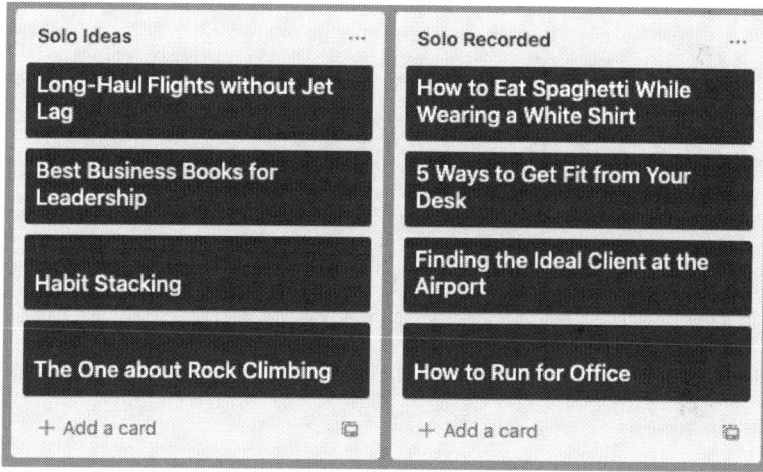

Fig. 11: Ideas and produced shows

The card itself can be used to keep notes and talking points that help you create a podcast episode that is not just interesting to your target market but useful to them too. And because you're familiar with the Profitable-Pod Method, it'll be easy to use them to "get in" with specific people in various ways, as well as just publishing them out.

The solo show card below includes a few notes on an episode with specific names who have become muses for this episode.

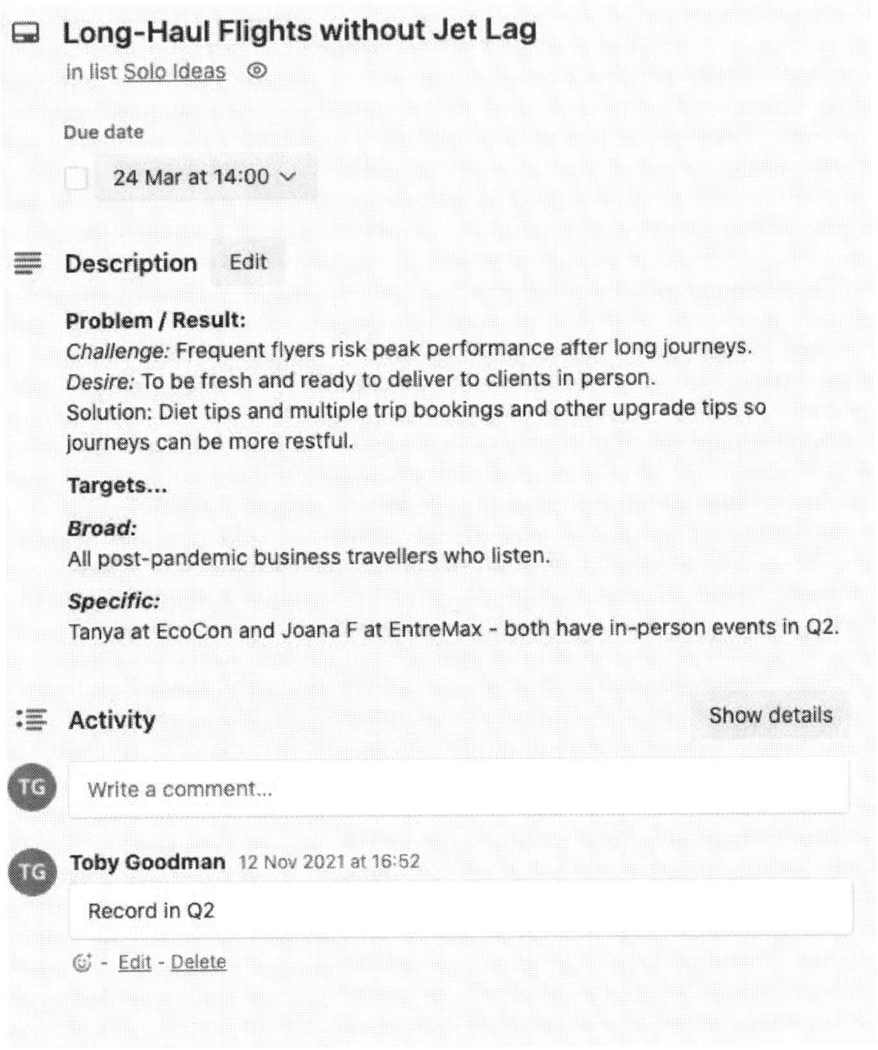

Long-Haul Flights without Jet Lag
in list Solo Ideas

Due date

24 Mar at 14:00

Description Edit

Problem / Result:
Challenge: Frequent flyers risk peak performance after long journeys.
Desire: To be fresh and ready to deliver to clients in person.
Solution: Diet tips and multiple trip bookings and other upgrade tips so journeys can be more restful.

Targets...

Broad:
All post-pandemic business travellers who listen.

Specific:
Tanya at EcoCon and Joana F at EntreMax - both have in-person events in Q2.

Activity   Show details

TG   Write a comment...

TG   **Toby Goodman** 12 Nov 2021 at 16:52

     Record in Q2

     - Edit - Delete

Fig. 12: Further notes on an idea

THE NARROW VIEW

- Set up your Kanban system (Trello or similar) to run your Pod-Lationship Cycle.
- Ask how you can connect your guest with other guests.
- Set reminders for follow-up connections on guest cards.
- Set broad and narrow targets for solo shows and don't forget to check them and see how they relate to new people in your network every few months.

11

SHOW ME THE MONEY (MORE WAYS OF GENERATING REVENUE)

The number-one question I get asked about making money from podcasting is: "When is it right to monetize?"

Hopefully you know by now the answer is never.

The truth is that the shortest path to the money is working the Profitable-Pod Method. Choose the right guests, for the right reasons, and then listen. This enables you to then leverage those relationships to grow your business. That's where the money in podcasting lies.

If you're really intentional about doing that, you may never need to "monetize." If you happen to grow your list at the same time, that's a fantastic secondary aim, but that's not the point of the Profitable-Pod Method. This chapter is more of a bonus section than anything. I've also written it here, so I never have to talk about it again! Only follow the advice in this chapter AFTER you have done everything else outlined in the book.

PERMISSION-BASED MARKETING

The term "permission-based marketing" is commonly attributed to Seth Godin. The idea is that you gain the prospect's permission to market to them, and they get to choose what they hear about. For most people, this leads to the now well-trodden path of Freebie / Opt In / Lead Magnet funnel. And that's how a lot of people choose to run a profitable pod. Which is totally cool. But remember, the primary goal is to leverage relationships with your guests.

If you grow your list using a lead magnet after they listen to you, that's a great bonus—providing you actually do something with that list and have things to offer them. But that's a completely different book.

Email Marketing

Let me be clear. You should send a short, polite email to your list every week letting them know who is on the show. In the PS section, let them know how they can get your support with links back to pages on your website. From a tech standpoint, you'll need a customer relationship management (CRM) system for that. As I said right in chapter 2, if you have a small list, some CRMs are free. In the meantime, here's one more way to generate specific leads from specific episodes that also requires an email automation tool.

Content Upgrades

A content upgrade is an offer linked to a specific episode, so while it will take some additional backend work on your website, it shouldn't take much additional work, especially if you have a team. If you don't, focus on the invitations and conversations.

For example, at the end of your episode, you can say, "If you enjoyed this episode, you can download an infographic about Ken's cheese-making process when you go to mywebsite.com/podcast/making-cheese-with-ken."

As long as a content upgrade in some way enhances the experience of the listener and is specific to the episode they listened to, the options are endless. You just need to make sure this is hooked up to deliver via your website. And this is where you get to segment your list so you can communicate with them in the appropriate way.

The classic showbiz idea behind all of this is "leave them wanting more." However, please don't get into this too soon, especially if you don't have a team member to help you.

Advertising

After I tell you that you shouldn't monetize, why am I talking about ads? Well, to start with, your podcast itself is an ad. It's an ad for how you show up in the world. It gets you in there with the right people by advertising who you are, what your expertise is, and how you serve your market. A professionally produced podcast is a great ad that has a much longer shelf life than anything on Facebook. And it's not just passive. You can use it like an electronic business card. Anytime you want to get in there with somebody new, you have this amazing asset you can send to them.

You should not launch with a sponsor. The truth is, your business is your sponsor. Your lead magnet will serve to be your very own ad. You have full control over it. You can say whatever you want, and you can change it whenever you want too! And that's an opportunity not to be missed.

However, you may find at some point that other people will want to sponsor your show. You will definitely be approached by a whole host of bizarre companies who will promise that they can get you sponsorship and will almost certainly want to know how many listeners you have. And you really do not want to play that game. No matter how many listeners you have, your stats will likely be wrong. That's because many of the platforms only count listeners when the episode is downloaded to a device. Almost certainly, most listeners will stream your show instead to save storage on their device. Those listens often get missed in the stats. If your show is positioned correctly and sponsorship makes sense for you, I encourage you to focus the value of sponsoring your show on the quality of guests and listeners rather than quantity of listeners.

> And you really do not want to play that game.

The next decent opportunity you might actually want to consider is a partner. Somebody who you believe serves the same target market as you but offers different things. Anybody who's talking to your market is likely to be excited by the opportunity to speak directly to a small, but highly engaged, audience in that market. And of course, you have one of those. Win.

The last kind of ad to consider is a dynamic ad. This ties back into "now offers." This is an advanced ad option, and you probably won't need or want to do it for maybe the first year. But it's a very powerful little trick, and it enables you to advertise your own time-sensitive stuff and partner with sponsors for a specific amount of time.

Let's say that on an original episode recording, you did a little piece promoting an upcoming event for either yourself or your guest. Once that event has passed, the ad is now effectively dead. But the beauty of dynamic ads is you can go back in time and change that excerpt to reflect something more evergreen. But there's another layer to this. You can sell ad space to sponsors for a finite period of time, say three months. At the end of the time, that sponsor has to decide whether they want to risk losing that space to a competitor. And the great part for you is, you can either sell them more time (and get paid again), sell a competitor their slot (and get paid again), or use it yourself and grow your business.

I would not recommend having more than three advertisers in your podcast. People tend to get tired of that stuff quickly (unless you're Adam Buxton, who has turned advertising on his podcast into an artform).

WHAT TO CHARGE

If you decide to go down the route of ads, this is an obvious question. The slightly less obvious answer starts with, "How much is one listener worth as a customer to the people buying the ad?"

If you've followed the advice in this book, your listeners are highly engaged and interested in a very specific result, which is something most advertisers can only dream of. Whatever you do, don't even attempt to engage with advertisers unless you're working with a professional production company and have AT LEAST six months' worth of shows under your belt. While every situation is unique, I believe that 99 percent of the time, "host-read ads" are more powerful. This means, as the host, you read the ads rather than a professional voiceover artist. In most cases, the less glossy and more direct the ad can be, the better. That's simply because your sponsor is paying for that transfer of trust from you to them. Just don't forget to have your production company time stamp your ad read, so you can remove and replace the ad if you need to.

> **THE NARROW VIEW**
>
> - Leverage relationships you create by using your podcast to grow your business.
> - Use permission-based marketing/opt ins/free downloads hosted on your website to harvest your relationships with listeners.
> - Send short weekly emails letting subscribers/followers of your podcast know about new episodes and other ways you may be able to help them.
> - When you're ready/if you have time resources, create additional episode-specific content upgrades to measure how each show performs.
> - Have your business sponsor your podcast.
> - Don't get caught out in "grow your podcast" vanity games. Grow your business, NOT your podcast.
> - If you are in a position to generate revenue for ads, use dynamic ad space that last a specific amount of time so it can be resold to the same or a new sponsor.

12
DON'T BE LIKE DAVE

The bloke who lives up the road from me makes more noise than everybody in the village. Every time there is a football game, a public holiday, or a sparrow's fart, Dave uses it as an excuse to have a party. While I have long accepted that some people will make a letter opening an excuse to celebrate, I absolutely can't stand the ones who force their need to be seen and heard on the innocent people who happen to be in the vicinity.

You see, Dave is an amateur DJ. And whenever he feels like it, he sets up his PA system on his front lawn and gives the village a party they haven't asked for. To my rough calculation, he has five friends and about 5000 enemies. If Dave were a podcaster (and he very well might be), he could never be my client. He'd be the kind of podcaster who just wants to be famous, or as the cool kids now say, "an influencer." By being too pushy and needing to be known as "the cool guy at number 32," he is loathed by everyone in a four-mile radius. By trying to be for everyone, Dave is not for anyone.

If "Double Decks Dave" really thought about it, he could easily find 100 local people who love Abba's greatest hits as much as he does, hire a local hall, and make good money for doing what he loves. But Dave clearly doesn't think about anyone but himself, so he is resigned to working a job to pay for his record collection that he unknowingly uses as a monthly torture device on his local community.

If you've come this far, I suspect you're more like me than Dave (God bless him), and I'd like to thank you for your time and attention. If you are an expert who is keen to make the right noise to the right people who will want to listen, a podcast is a great way to do it, especially when you move away from a broadcaster, mud-at-the-wall approach and use the narrowcast strategy outlined in this book.

When I go to do the weekly shopping and don't take a list with me, I invariably spend too much money, too much time, and I buy the wrong stuff. But when I take my list, while I allow myself the odd addition, I go home with everything we need. Now we're coming to the end of the book, it's time I gave you a narrow podcasting list. Addressing the major points of the Profitable-Pod Method, the list ensures you continue to keep your focus narrow and avoid the vanity trap so that you can create a podcast to serve the business you are building.

ULTIMATE PROFITABLE POD CHECKLIST

Show Planning & Positioning

I understand that the Profitable-Pod Method …

- ✓ … is not the same as TV or radio, and I'm not hosting a shiny-floor TV show for "the people."
- ✓ … needs my speaking tone to remain relaxed and personal so it connects to my guests and listeners.

I have …

- ✓ … a narrow target market and have positioned my podcast to appeal to them and/or the people they have relationships with.
- ✓ … a strong show title/tagline that immediately explains who my show is for, and who it isn't for, so that I'm not leaving any listeners who come across my podcast in any doubt about what they are in for if they press play.
- ✓ … created at least one freebie (guide/quiz) that speaks to the needs/desires of my ideal clients.
- ✓ … great-looking artwork that creates a strong first impression, showing my guests, people in my network, and listeners that I take pride in my work.

Gear / Tech Stack / Production

My computer is in good condition, and I have access to a stable high speed internet connection.

I have (or plan to acquire soon) ...

- ✓ ... a paid-for, legitimate hosting package so I can be confident I have full ownership of my own podcast content and my show can't be hijacked.
- ✓ ... quality recording equipment that's set up in a quiet environment.
- ✓ ... the help of professional production services.
- ✓ ... an automated email/CRM system linked to my website that can deliver my freebie, email blasts, and any additional content upgrades.
- ✓ ... a system to run my Pod-Lationship Cycle (e.g., Trello or similar).
- ✓ ... a system (Canva, Wavve, or similar) to easily create social media assets.

My website ...

- ✓ ... can host my podcast with blog posts set up for each individual episode.
- ✓ ... is easy to navigate and my podcast is easily found.
- ✓ ... helps guests book in for a recording, while simultaneously enabling me to collect information from guests to create show notes, episode artwork, and additional social media assets.
- ✓ ... hosts a lead magnet (a short value based freebie) that people who don't know me can download and identify themselves to me as interested in me so that I can follow up and find out more about them.

Pre-launch Production

I am clear on the possible listening positions (sensitive, defensive, critical, open, etc.) of ...

- ✓ ... my various guests
- ✓ ... my ideal listeners

I have …

- ✓ … written and recorded my intro, outro, lead magnet, and teaser script elements.
- ✓ … made considered invites to guests, taking time to clarify
    - a) my own good/better/best outcomes for each show.
    - b) why they would want to be on my podcast.
- ✓ … planned my first ten episodes for launch.
- ✓ … carefully created five solo episodes that showcase my expertise with specific people in mind so I can use the shows as a tool to generate interest in my podcast launch from day one.
- ✓ booked/recorded at least five guest episodes.

Every guest I invite is one or all of the following …

- ✓ … an existing client/prospect
- ✓ … powerful referrer
- ✓ … partner

I will …

- ✓ … be ready to publish weekly episodes seven days after my launch.
- ✓ … maintain a steady stream of invites and solo shows so I'm never stuck for recordings and never need to create under pressure.
- ✓ … plan thoughtful questions with quality answers in mind that will serve to
    - a) enhance my reputation with my guest.
    - b) create a sense of inclusivity for my listeners.

My production team helps me …

- ✓ … set up my recording gear to get best results and edit my audio.
- ✓ … match music and sound to my show.
- ✓ … write SEO show notes with pullout quotes.
- ✓ … upload my podcast to my hosting platform.

- ✓ … by communicating with me clearly.
- ✓ … create a podcast art and audiogram system that I can easily drag into graphic design tools I use.

Pod-Lationship Cycle/Follow Up

- ✓ I have set up a pool of pod-prospect guests.
- ✓ I'm making frequent invitations to the right people who fit one or more of the essential three criteria in the Profitable-Pod Method (potential client, referrer, partner).
- ✓ I have a keep-in-touch strategy that …

    a) builds my credibility, likeability with my guests.

    b) builds a wider network that makes it easy for my guests to share the podcast.

    c) gives me a list of podcast guest alumni to keep in regular contact with.

- ✓ I separate the network I built through my guest interviews from my mass email lists and never opt them into my "out there" stuff.
- ✓ I am always on the lookout for ways I can serve my network.

FROM POD-CURIOUS TO PROFITABLE POD-LATIONSHIPS

Thanks so much for investing in and allowing me to show you the narrow path to a profitable podcast. We've covered lots in this book from early planning and preproduction, all the way through to your Pod-Lationship Cycle to help you build connections leading to the business you want. If you're the kind of person who has always hoped that your work would speak for itself, narrow podcasting is perfect for you because you're giving your business a voice.

You've come this far, so I already know you're passionate about what you do. But if people don't know that, you're never going to get the success you deserve. It's simply naive to put people before profits, but putting your target market before profits absolutely isn't. In fact, your target market is the first thing you should consider.

You and your deep knowledge of the needs and desires of your target market are unique. So, when it comes to all things narrow podcasting, please adapt what you've discovered about the Profitable-Pod Method to suit your voice and your people. And please send me a note and let me know how it's working for you.

When it comes to creating your podcast, expect to be overwhelmed. There are lots of moving parts. Honestly, it took writing this book to realize just how many! And while I know this book shows you everything you need to know about the method and how to get started, I couldn't cover everything. There are simply too many variables. That said, you should take comfort. You've done more than most. Please also be assured that I've never met a successful business owner with a podcast that serves them, who doesn't get support in some way.

Now we've come to the end of the book, but this doesn't have to be goodbye. Whenever you're ready, I (along with my amazing team of podcast professionals) am here to support you on your narrow and profitable podcast journey. Because nothing says, "I don't care about my business" like a podcast that sounds like a voicemail. Your time is best spent doing what only you can do.

From additional resources, tips, hosting deals, the Profitable-Pod Method online course, braodcast quality production, and even info on working with me privately including how to become a Profitable Pod Professional. You'll find more narrow podcasting gold over on my website, https://narrowpodcasting.com.

To your pod-success,
Toby

THANKS AND ACKNOWLEDGMENTS

Thanks to Sarah, Stanley, and Harry for your love and support. I'm a very lucky guy!

Kate! Your love and trust are appreciated more than words can express.

Tayler, Iris, and Lee. Legends, all three of you.

James Eager. I'm so proud of what you have done since we decided to take control!

John Cottrell. Thanks for lighting the fire and showing me the doors.

Julian Treasure. You helped me connect the value and articulated what I never could before.

Keith Tippett for teaching me how to listen. Still learning. RIP.

Dr. Kathy Sullivan. My mind is blown! Working with you is an honor and an inspiration.

Matthew Kimberley. Thanks for having me at *Book Yourself Solid* and introducing me to Michael's work.

Brady "selfless generosity" Sadler. Keep up the good work at Double Elvis.

Leo Hornak. Your pod skills are an inspiration.

James Langridge. Thanks for your time, patience, and considerable input.

Lorna Partington, thanks for all your time, energy, and direction.

Todd Cochrane, Mike, Makenzie and the Blubrry Team!

Thanks to every musician I have ever worked with.

I stand on the shoulders of giants, including but not even close to being limited to:

Billy Ward, Simon Edgoose, Darren Williams, Mark J. Silverman,

Helen Appleby, Karen Goldfinger Baker, Chris Colbert, Dayna Steele,

Shweta Jhajharia, Marsha Shandur, Kymberlee Jay, Jos Willard, Liz Scully, Andrew Lock, Dov Gordon, Rachel Allen, Suzy Bastone, Juliet Corbett, Stephanie Cartin, Murtaza Versi, Dr. Marty Fletcher, Gordon Firemark, Howard Cooper, Blake Ezra, Dr. Anna Symonds, Saranne Rothberg, Callan Leader, Michael Port, James Schramko, Taki Moore, Todd Herman, Dean Jackson, Derek Sivers.

ABOUT THE AUTHOR

As a gun-for-hire musician, Toby Goodman has traveled the world playing for a Rolling Stone, a Bee Gee, a Spice Girl, and more TV talent show winners (and losers) than the press can remember. But he has another life. One in which he's a coach and consultant to TED speakers, CEOs, and even an astronaut. He helps them to harness the power of podcasting to grow their businesses, influence, and authority. He also heads a US-founded podcast production company with a global team and client base. All while being a husband and father with two small children in the leafy suburbs of London.

To find out how he can help you grow your business with a podcast, head over to: narrowpodcasting.com.

Made in the USA
Columbia, SC
20 May 2022